Material Modeling in Finite Element Analysis

Material Modeling in Finite Element Analysis

Z. Yang

CRC Press
Taylor & Francis Group
Boca Raton London New York

CRC Press is an imprint of the
Taylor & Francis Group, an **informa** business

CRC Press
Taylor & Francis Group
6000 Broken Sound Parkway NW, Suite 300
Boca Raton, FL 33487-2742

© 2020 by Taylor & Francis Group, LLC
CRC Press is an imprint of Taylor & Francis Group, an Informa business

No claim to original U.S. Government works

Printed on acid-free paper

International Standard Book Number-13: 978-0-367-35320-9 (Hardback)

This book contains information obtained from authentic and highly regarded sources. Reasonable efforts have been made to publish reliable data and information, but the author and publisher cannot assume responsibility for the validity of all materials or the consequences of their use. The authors and publishers have attempted to trace the copyright holders of all material reproduced in this publication and apologize to copyright holders if permission to publish in this form has not been obtained. If any copyright material has not been acknowledged please write and let us know so we may rectify in any future reprint.

Except as permitted under U.S. Copyright Law, no part of this book may be reprinted, reproduced, transmitted, or utilized in any form by any electronic, mechanical, or other means, now known or hereafter invented, including photocopying, microfilming, and recording, or in any information storage or retrieval system, without written permission from the publishers.

For permission to photocopy or use material electronically from this work, please access www.copyright.com (http://www.copyright.com/) or contact the Copyright Clearance Center, Inc. (CCC), 222 Rosewood Drive, Danvers, MA 01923, 978-750-8400. CCC is a not-for-profit organization that provides licenses and registration for a variety of users. For organizations that have been granted a photocopy license by the CCC, a separate system of payment has been arranged.

Trademark Notice: Product or corporate names may be trademarks or registered trademarks, and are used only for identification and explanation without intent to infringe.

Visit the Taylor & Francis Web site at
http://www.taylorandfrancis.com

and the CRC Press Web site at
http://www.crcpress.com

Contents

Preface xv
Author xvii

1. Introduction 1

Part I Metal

2. Structure and Material Properties of Metal 7
   2.1 Structure of Metal 7
   2.2 Elasticity and Plasticity of Metal 7
   Reference 10

3. Some Plastic Material Models of Metals and Definition
   of Their Parameters 11
   3.1 Introduction of Plasticity 11
   3.2 Isotropic Hardening Models and Definition of Material
        Parameters for 304 Stainless Steel 12
        3.2.1 Multilinear Isotropic Hardening (MISO) 12
        3.2.2 Voce Law Nonlinear Isotropic Hardening 16
   3.3 Nonlinear Kinematic Hardening 19
   3.4 Summary 21
   References 21

4. Simulation of Metal Forming 23
   4.1 Introduction of Metal Forming 23
   4.2 Simulation of Forming of a Sheet 24
        4.2.1 Finite Element Model 24
        4.2.2 Material Properties 26
        4.2.3 Contact Definition 26
        4.2.4 Loadings and Boundary Conditions 26
        4.2.5 Solution Setting 26
        4.2.6 Results 27
        4.2.7 Summary 27
   References 30

5. Simulation of Ratcheting 31
   5.1 Introduction of Ratcheting 31
   5.2 Simulation of Ratcheting in a Notched Bar 31
        5.2.1 Finite Element Model 31

	5.2.2	Material Properties 33
	5.2.3	Loadings and Boundary Conditions 33
	5.2.4	Results 33
	5.2.5	Summary 36
References 36

6. Influence of Temperature on Material Properties 37
6.1 Temperature Dependency of Material Properties 37
6.2 Simulation of Combustion Chamber under Different Temperatures 38
    6.2.1 Finite Element Model 38
    6.2.2 Material Properties 38
    6.2.3 Loadings and Boundary Conditions 40
    6.2.4 Results 40
    6.2.5 Discussion 40
    6.2.6 Summary 45
References 45

7. Simulation of Creep 47
7.1 Introduction of Creep 47
    7.1.1 Creep 47
    7.1.2 Creep Constitutive Law 47
    7.1.3 Subroutine UserCreep 49
7.2 Simulation of Creep of a Bolt under Pretension 53
    7.2.1 Finite Element Model 53
    7.2.2 Material Properties 53
    7.2.3 Loadings and Boundary Conditions 54
    7.2.4 Solution Setting 55
    7.2.5 Results 55
    7.2.6 Discussion 55
    7.2.7 Summary 57
References 57

Part II Polymers

8. Structure and Features of Polymer 61
8.1 Structure of Polymer 61
8.2 Features of Polymer 61
References 62

9. Hyperelasticity 63
9.1 Some Widely Used Hyperelastic Models 63
    9.1.1 Neo-Hookean Model 63

	9.1.2	Mooney–Rivlin Model	63
	9.1.3	Yeoh Model	64
	9.1.4	Polynomial Model	64
	9.1.5	Gent Model	64
	9.1.6	Ogden Model	64
	9.1.7	Arruda–Boyce Model	65
9.2	Stability Discussion	66
9.3	Curve-fitting of Material Parameters from Experimental Data	67
9.4	Simulation of a Rubber Rod under Compression	69
	9.4.1	Finite Element Model	69
	9.4.2	Material Parameters	69
	9.4.3	Loadings and Boundary Conditions	70
	9.4.4	Results	70
	9.4.5	Discussion	70
	9.4.6	Summary	71
9.5	Simulation of Breast Implant in ANSYS	72
	9.5.1	Finite Element Model	73
	9.5.2	Material Models	74
	9.5.3	Loading and Solution Setting	75
	9.5.4	Results	75
	9.5.5	Discussion	75
	9.5.6	Summary	77
References	77

10. Viscoelasticity of Polymers ... 79

10.1	Viscoelasticity of Polymers	79
10.2	Linear Viscoelastic Models	79
	10.2.1 Maxwell Model	80
	10.2.2 Kelvin–Voigt Model	80
	10.2.3 Burgers Model	81
	10.2.4 Generalized Maxwell Model	81
10.3	Viscoplasticity Models	83
10.4	Simulation of Viscoelasticity of Liver Soft Tissues	84
	10.4.1 Finite Element Model	84
	10.4.2 Material Properties	85
	10.4.3 Contact Definition	85
	10.4.4 Loadings and Boundary Conditions	85
	10.4.5 Results	86
	10.4.6 Discussion	87
	10.4.7 Summary	87
References	87

11. Mullins Effect ... 89

11.1	Introduction of Mullins Effect	89

11.2 Ogden–Roxburgh Mullins Effect Model 90
11.3 Simulation of a Rubber Tire with the Mullins Effect 90
    11.3.1 Finite Element Model 91
    11.3.2 Material Properties 91
    11.3.3 Loadings and Boundary Conditions 91
    11.3.4 Results 92
    11.3.5 Discussion 92
    11.3.6 Summary 98
References 99

12. Usermat for Hyperelastic Materials 101
12.1 Introduction of Subroutine UserHyper 101
12.2 Simulation of Gent Hyperelasticity 101
    12.2.1 Subroutine UserHyper for Gent Material 102
    12.2.2 Validation 103
    12.2.3 Summary 104
References 104

Part III Soil

13. Soil Introduction 107
13.1 Soil Structure 107
13.2 Soil Parameters 107
References 108

14. Cam Clay Model 109
14.1 Introduction of Modified Cam Clay Model 109
14.2 Cam Clay Model in ANSYS 111
    14.2.1 Elastic Component 111
    14.2.2 Plastic Component 112
14.3 Simulation of a Tower on the Ground by Cam Clay Model 113
    14.3.1 Finite Element Model 113
    14.3.2 Material Properties 113
    14.3.3 Contact Definition 114
    14.3.4 Loadings and Boundary Conditions 114
    14.3.5 Results 114
    14.3.6 Discussion 118
    14.3.7 Summary 118
References 118

15. Drucker–Prager Model 119
15.1 Introduction of Drucker–Prager Model 119
15.2 Study of a Soil–Arch Interaction 120

	15.2.1 Finite Element Model 120
	15.2.2 Material Properties 121
	15.2.3 Boundary Conditions and Loadings 122
	15.2.4 Results 122
	15.2.5 Discussion 124
	15.2.6 Summary 124
	References 125

16. Mohr–Coulomb Model 127
16.1 Introduction of Mohr–Coulomb Model 127
16.2 Mohr–Coulomb Model in ANSYS 130
16.3 Study of Slope Stability 131
    16.3.1 Finite Element Model 131
    16.3.2 Material Properties 132
    16.3.3 Loadings and Boundary Conditions 133
    16.3.4 Results 133
    16.3.5 Discussion 135
    16.3.6 Summary 136
References 136

17. Jointed Rock Model 139
17.1 Jointed Rock Model 139
17.2 Definition of the Jointed Rock Model in ANSYS 140
    17.2.1 Defining the Base Material 140
    17.2.2 Defining the Joints 141
17.3 Simulation of Tunnel Excavation 141
    17.3.1 Finite Element Model 141
    17.3.2 Material Properties 141
    17.3.3 Loadings and Boundary Conditions 142
    17.3.4 Solution 142
    17.3.5 Results 142
    17.3.6 Discussion 145
    17.3.7 Summary 146
References 146

18. Consolidation of Soils 149
18.1 Consolidation of Soils 149
18.2 Modeling Porous Media in ANSYS 149
18.3 Simulation of Consolidation of Three-Well Zone 150
    18.3.1 Finite Element Model 150
    18.3.2 Material Properties 150
    18.3.3 Boundary Conditions and Loadings 152
    18.3.4 Solutions 152
    18.3.5 Results 152

		18.3.6	Discussion	154
		18.3.7	Summary	155
	References			156

Part IV Modern Materials

19. Composite Materials 159
- 19.1 Introduction of Composite Materials 159
- 19.2 Modeling Composite in ANSYS 161
  - 19.2.1 Modeling Composite by Command SECTYPE 161
  - 19.2.2 Modeling a Composite by Anisotropic Model 162
- 19.3 Simulation of Composite Structure in Failure Test 163
  - 19.3.1 Finite Element Model 163
  - 19.3.2 Material Properties 163
  - 19.3.3 Boundary Conditions and Loadings 166
  - 19.3.4 Results 166
  - 19.3.5 Discussion 169
  - 19.3.6 Summary 169
- 19.4 Simulation of Crack Growth in Single Leg Bending Problem 170
  - 19.4.1 Finite Element Model 171
  - 19.4.2 Material Properties 171
  - 19.4.3 Crack Definition 172
  - 19.4.4 Boundary Conditions and Loadings 172
  - 19.4.5 Results 172
  - 19.4.6 Discussion 173
  - 19.4.7 Summary 174
- References 174

20. Functionally Graded Materials 175
- 20.1 Introduction of Functionally Graded Materials 175
- 20.2 Material Model of Functionally Graded Materials 175
- 20.3 Simulation of a Spur Gear Fabricated Using Functionally Graded Materials 176
  - 20.3.1 Finite Element Model 176
  - 20.3.2 Material Properties 177
  - 20.3.3 Loadings and Boundary Conditions 178
  - 20.3.4 Results 179
  - 20.3.5 Discussion 179
  - 20.3.6 Summary 179
- References 182

21. Shape Memory Alloys 183
- 21.1 Structure of SMAs and Various Material Models 183

   21.1.1 Structure of SMAs 183
      21.1.1.1 Superelasticity.. 183
      21.1.1.2 Shape Memory Effect.................................. 185
   21.1.2 Various SMA Material Models... 186
      21.1.2.1 SMA Model for Superelasticity 186
      21.1.2.2 SMA Model with Shape Memory Effect........... 188
   21.1.3 Definition of Material Parameters 189
      21.1.3.1 SMAs with Superelasticity......................... 189
      21.1.3.2 SMAs with Shape Memory Effect.................... 190
 21.2 Simulation of Orthodontic Wire 191
   21.2.1 Finite Element Model 191
   21.2.2 Material Properties 192
   21.2.3 Loadings and Boundary Conditions 193
   21.2.4 Results 193
   21.2.5 Discussion 195
   21.2.6 Summary.. 196
 21.3 Simulation of a Vacuum-Tight Shape Memory Flange................. 196
   21.3.1 Finite Element Model 196
   21.3.2 Material Properties 199
   21.3.3 Contact.. 199
   21.3.4 Loadings and Boundary Conditions 199
   21.3.5 Solutions... 199
   21.3.6 Results 200
   21.3.7 Discussion 200
   21.3.8 Summary..204
 References 204

22. Simulation of Piezoelectricity..207
 22.1 Introduction of Piezoelectricity 207
 22.2 Structures and Mechanical Behaviors of Piezoelectric
     Materials...208
 22.3 Constitutive Equation of Piezoelectricity 209
 22.4 Simulation of Piezoelectric Accelerometer.................................... 211
   22.4.1 Finite Element Model 211
   22.4.2 Material Properties 212
   22.4.3 Boundary Conditions and Loadings 213
   22.4.4 Results 214
   22.4.5 Discussion 215
   22.4.6 Summary.. 217
 References 218

23. Nanomaterials 219
 23.1 Introduction of Nano... 219
 23.2 Determination of Young's Modulus of Fe Particles 221
   23.2.1 Experiment.. 221

	23.2.2	Finite Element Model	221
	23.2.3	Material Properties	222
	23.2.4	Boundary Conditions and Loadings	222
	23.2.5	Solution	222
	23.2.6	Results	223
	23.2.7	Discussion	223
	23.2.8	Summary	224
References			224

Part V Retrospective

24 Retrospective ... 227

Appendix 1: Input File of Curve-Fitting of the Chaboche Model in Section 3.3 ... 231

Appendix 2: Input File of the Forming Process Model in Section 4.2 ... 233

Appendix 3: Input File of the Ratcheting Model in Section 5.2 ... 237

Appendix 4: Input File of the Combustion Chamber Model in Section 6.2 ... 239

Appendix 5: Input File of the Bolt Model under Pretension in Section 7.2 ... 243

Appendix 6: Input File of Curve-Fitting of the Ogden Model in Section 9.3 ... 245

Appendix 7: Input File of the Rubber Rod Model under Compression in Section 9.4 ... 247

Appendix 8: Input File of the Liver Soft Tissue Model in Section 10.4 ... 251

Appendix 9: Input File of the Rubber Tire Damage Model in Section 11.3 ... 255

Appendix 10: Input File of UserHyper in Section 12.2 ... 259

Appendix 11: Input File of the Tower Subsidence Model in Section 14.3 ... 261

Appendix 12: Input File of the Soil–Arch Interaction Model in Section 15.2 ... 265

Appendix 13: Input File of the Slope Stability Model in Section 16.3 ... 269

Contents xiii

Appendix 14: Input File of the Tunnel Excavation Model in Section 17.3 271

Appendix 15: Input File of the Settlement Model in Section 18.3 273

Appendix 16: Input File of the Composite Damage Model in Section 19.3 277

Appendix 17: Input File of the SLB Model in Section 19.4 281

Appendix 18: Input File of the Spur Gear Model with FGM in Section 20.3 285

Appendix 19: Input File of the Orthodontic Wire Model in Section 21.2 287

Appendix 20: Input File of the Vacuum Tight Shape Memory Flange Model in Section 21.3 289

Appendix 21: Input File of the Piezoelectric Microaccelerometer Model in Section 22.4 293

Appendix 22: Input File of the Contact Model in Section 23.2 303

Index 305

Preface

Twenty years ago, I had an educational experience that has defined the direction of my life. I realized that the vibration of a pipe in a chemical plant would dramatically reduce after the application of constraints following the first vibration mode provided by finite element analysis. Therefore, when I came to the United States to pursue my PhD at the University of Pittsburgh, I enthusiastically selected finite element analysis as my research direction. After my graduation, I worked with Frank Marx at Westinghouse Electric Corporation to simulate the vibration of a control room of a nuclear plant under an earthquake. Later, I developed a finite element program to mimic the water flow in rocks for the National Energy Technology Laboratory. I have spent the past 12 years simultaneously working as a software engineer in the Research and Development Department and conducting finite element analyses in different areas, including the mechanical, civil, and biomedical fields. I have completed more than 20 projects; these projects have provided me with insights into finite element analysis, especially in material modeling. In this book, I try to summarize my knowledge of experiences with material modeling to give the readers a deeper understanding of the various materials required to build correct material models and select the proper solution setting for practical problems.

During the writing and publication of this book, I received help from many of my friends, including Frank Marx, Dr. J.S. Lin, Dr. Zhi-Hong Mao, Dr. Adi Adumitroaie, and Ronna Edelstein. Without their help, I could not have finished such a book. I also greatly appreciate the staff at CRC Press, especially Marc Gutierrez, for their assistance in publishing this book. Finally, I thank my wife, Peng Tang, and two children, Tom and Amy, for their constant support and encouragement.

Author

Z. Yang received his PhD degree in Mechanical Engineering from the University of Pittsburgh in 2004. Since 2005, he has worked for big companies and the national labs such as National Energy Technology Laboratory (NETL). He has been in the field of finite element analysis over 17 years, and gained much experience, especially in the aspect of material modeling. So far, he has published 12 journal papers. In 2019, CRC Press published his first book, *Finite Element Analysis for Biomedical Engineering Applications*.

1
Introduction

We live in a modern world of buildings, cars, airplanes, and medical advances, including coronary angioplasty. Buildings require structural analysis from civil engineers, while the manufacture of cars and airplanes depends upon strong mechanical design. Biomedical engineering plays a significant role in coronary surgeries. All these fields demand an understanding of the stress state of the structures. With the development of computer technology, finite element analysis has been widely used in these fields. Specifically, since the 1970s, many commercial finite element software companies have evolved: ANSYS, ABAQUS, COMSOL, ADINA, LS-DYNA, and MARC. Among these, ANSYS emerges as the leader.

One of the key processes in finite element analysis is to build material models. The structural analysis of a building is associated with soils, the manufacture of a car involves metals, the airplane has many components made of composites to ensure both high strength and low weight, and the stent in the coronary angioplasty is made of shape memory alloys. When we create finite element models to conduct stress analysis for their designs, we must adopt a variety of material models corresponding to different materials, because under the same geometry and loadings and boundary conditions, different materials may have unique behaviors. For example, it is hard to pull a steel rod, but the same force can easily deform a rubber rod. If we pull the metal rod to plasticity, the plastic strains remain after the force is gone. However, when the metal rod is made of shape memory alloys, the remaining transformation strains can be removed with the rise of temperature. Unlike these materials, soil has little resistance to tension. The purpose of this book is to acknowledge that different materials have unique features and to present readers with comprehensive views of various material models. By providing practical examples, the book will enable readers to better understand the material models, how to select appropriate material models, and how to define the correct solution set in the finite element analysis of complicated engineering problems. As discussed above, ANSYS is the leading software in the field of finite element analysis. Thus, the examples in the book are provided in the form of ANSYS input files, which makes it convenient for readers to learn and practice these examples.

The book is composed of four main parts. After the introductory chapter, the first part focuses on metals and alloys. Chapter 2 introduces the structure of alloys and mechanical features. Some plastic models such as the bilinear isotropic hardening model, the multilinear isotropic hardening model, the

Voce law nonlinear isotropic hardening model, and the Chaboche model, along with examples showing how to determine material parameters from the experimental data, are presented in Chapter 3. Chapter 4 discusses the application of plastic models for simulation of the forming process, including one example of a sheet that is deformed between two dies. Ratcheting is one unique feature of plastic metals under cyclic loading. It is reproduced on a notched rod with the Chaboche kinematic hardening material model under cyclic loading in Chapter 5. As the material properties of metals are temperature-dependent, Chapter 6 examines the influence of temperature on a combustion chamber. The last chapter in Part I develops the creep subroutine and applies it to simulate the creep behavior of the metal under pretension.

Part II discusses polymers. After Chapter 8 depicts the structure and material properties of polymers, Chapter 9 presents some hyperelastic material models, including two examples: the large deformation of a rubber rod and the deformation of a breast after breast surgery. The viscoelasticity of polymers and its application for liver soft tissues are discussed in Chapter 10. The stress responses of elastomers always experience softening during the first few loading cycles. That is regarded as a damage accumulation in the material, which refers to the Mullins effect of elastomers. Chapter 11 focuses on this and includes an example of a rubber tire. Userhyper is available in ANSYS for the customers to create their own hyperelastic models. Chapter 12 presents one example of Userhyper to reproduce the Gent model.

Soils are the topic of Part III. Chapter 13 introduces the structure and various classifications of soils. Four major material models of soils – the Cam Clay model, Drucker–Prager model, Mohr–Coulomb model, and jointed rock model – are discussed in Chapters 14 to 17, respectively. These chapters also cover some practical problems: a tower on the ground, soil–arch interaction, the stability of a slope, and a tunnel excavation. Since soils are composed of rocks, water, and air, Chapter 18 focuses on the consolidation of soils and its application for simulation of consolidation of rocks with three wells.

Part IV highlights modern materials. One widely used modern material is composite, which was developed in the 1950s due to airplane design needing high strength and low weight. Structure and material properties of composites are presented in the first part of Chapter 19, followed by an application on flight-qualification testing and crack growth in single-leg bending problem.

Chapter 20 examines functionally graded materials and their simulation in ANSYS using TBFIELD technology.

The unique features of shape memory alloys (SMAs), superelasticity and shape memory effect, are discussed in Chapter 21. The chapter includes two examples to demonstrate SMAs' applications: (1) orthodontic wire using the superelasticity feature and (2) a vacuum-tight shape memory flange using shape memory effect.

Introduction 3

Chapter 22 reviews the structure and mechanical behavior of piezoelectric materials. This is followed by the simulation of a thin film piezoelectric microaccelerometer using the piezoelectric material model in ANSYS.

Nanoscale materials refer to a group of substances with at least one dimension less than approximately 100 nm, which attract more and more interest. The first part of Chapter 23 introduces the nanoscale materials; in the second part, Young's modulus of nano Fe particles is determined from the experimental data of a ball made of Fe particles using the optimization algorithm in ANSYS.

Chapter 24 reviews the features of metals/alloys, polymers, soils, and modern materials, and discusses the relation between material properties and the structures as well as the relation between the temperature and the structure. It also examines solution control for various materials, including anisotropic materials with symmetrical conditions.

Part I

Metal

Metal plays a dominant role in modern industry, especially in the manufacturing of cars, airplanes, and ships. Its mechanical design requires a clear understanding of its stress-and-strain state. Therefore, Part I focuses on many metal material models implemented in ANSYS.

Chapter 2 introduces the structure and material properties of metal. Some plastic models, including the material parameters determined from the experimental data, are discussed in Chapter 3. Chapters 4 and 5 simulate metal forming and ratcheting, respectively. The influence of temperature on the material properties is analyzed in Chapter 6, and creep is simulated by both ANSYS and user subroutine in Chapter 7.

2
Structure and Material Properties of Metal

The material properties of metal are closely linked with the structure of metal. Thus, this chapter first presents the structure of metal and then gives the material properties of metal.

2.1 Structure of Metal

A metal is a material in which the atoms are joined together by metallic bonds (Figure 2.1). The delocalized electrons and the strong interaction forces between the positive atom nuclei give the metal features such as good thermal and electrical conduction.

Assuming the material's atoms as a perfect crystal, the bond energy is expressed as [1]

$$U = \frac{-ACe^2}{r} + \frac{B}{r^n} \qquad (2.1)$$

where $A = 1.747558$, $C = 8.988 \times 10^9$ N − m²/Coul².

As shown in Figure 2.2, the curve is more flattened out at larger separation distances. When the distance of atoms is less than r_0, the forces between the atoms are repulsive. When the distance of atoms is greater than r_0, the forces become attractive.

Anharmonicity of the bond energy function explains why materials expand when the temperature increases (Figure 2.3). As the internal energy rises due to the addition of heat, the system oscillates between the new positions marking A' and B' from the original positions A and B. Since the curve is not shaped like a sine curve, the new separation distance is longer than the previous one. Thus, the material has expanded, thereby causing the thermal strain.

2.2 Elasticity and Plasticity of Metal

When a force is loaded onto the metal, the layers of atoms start to roll over each other. If the force is released, the layers of atoms fall back to their original positions. Then, the metal is regarded as elastic (Figure 2.4a).

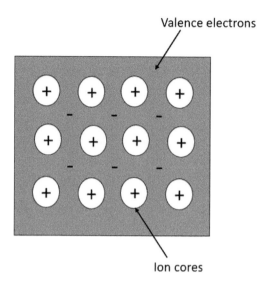

FIGURE 2.1
Atoms are held together by metallic bonds in a metal.

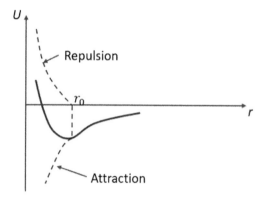

FIGURE 2.2
Bond energy function.

If the force is big enough that the atoms cannot fall back to their starting positions after the force is released, the metal is permanently changed (Figure 2.4b).

Based on the above statements, the uniaxial tension tests of the metal show that when the stresses of the metal are below the yield stress σ_Y, the metal is elastic. When the stresses of the metal are higher than the yield stress σ_Y, the metal yields (Figure 2.5). Yield driven by shearing stresses slides one plane along another (Figure 2.6). The plastic deformation due to yield is a viscous

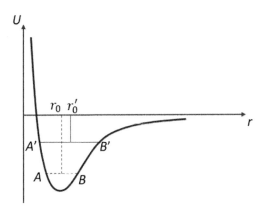

FIGURE 2.3
Anharmonicity of the bond energy function [1].

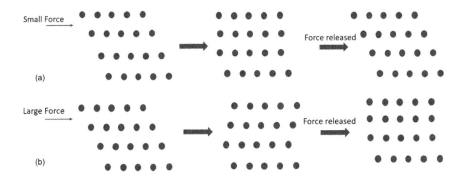

FIGURE 2.4
Illustration of a metal in elastic and plastic stages. (a) elastic stage, (b) plastic stage.

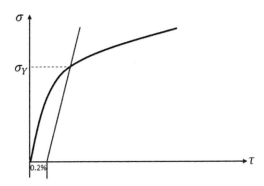

FIGURE 2.5
Stress–strain curve of the metal.

FIGURE 2.6
Sliding is induced by shear stresses.

flow process. Like flow in liquids, plastic flow has no volume change with Poisson's ratio $v=0.5$.

Normally, if the material is in the elastic stage, it is not likely to fail. This is not true for brittle materials like ceramics because ceramics fracture before they yield. However, for most of the structural materials, no damage occurs before yield. Thus, in the structural design, the materials required within the elastic stage use a safety factor. However, the plasticity of the metal has its application for industry such as metal forming. Therefore, some plastic models are introduced in Chapter 3.

Reference

1. Roylance, D., *Mechanical Properties of Materials*, MIT, 2008.

3

Some Plastic Material Models of Metals and Definition of Their Parameters

After introducing the plasticity theory, Chapter 3 presents some isotropic hardening models and kinematic hardening models; it also includes examples to determine the material parameters of these models from the curve-fitting of the experimental data.

3.1 Introduction of Plasticity

In the plasticity, the total strain is composed of the elastic strain ε^{el} and plastic strain ε^{pl}:

$$\varepsilon = \varepsilon^{el} + \varepsilon^{pl} \tag{3.1}$$

The stress is determined by the elastic strain ε^{el}

$$\sigma = D\varepsilon^{el} \tag{3.2}$$

where D is the stiffness matrix.

The general yield function is expressed as

$$f(\sigma,\xi) = 0 \tag{3.3}$$

where ξ refers to a set of internal variables.

When $f(\sigma,\xi) < 0$, it is inside the yield surface and in an elastic state. No stresses exist outside the yield surface. The stresses are either on or inside the yield surface.

Hardening exists when the increase in stress leads to an increase in plastic strain. There are two types of hardening: (1) isotropic hardening of the yield surface (see Figure 3.1) and (2) kinematic hardening of the yield surface (see Figure 3.2). The isotropic hardening changes the size of the yield surface but keeps the shape of the yield surface. It can model the behavior of the metals under monotonic loading. On the other hand, kinematic hardening changes both the size and the shape of the yield surface. It is observed in cyclic loading. Sometimes, both hardening rules are used together to present the combined hardening model.

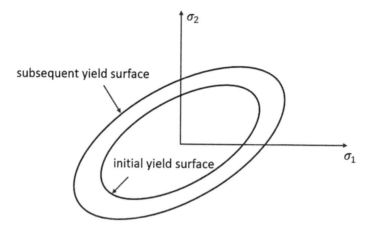

FIGURE 3.1
Isotropic hardening.

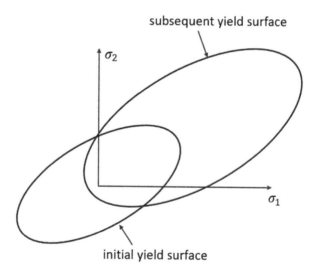

FIGURE 3.2
Kinematic hardening.

3.2 Isotropic Hardening Models and Definition of Material Parameters for 304 Stainless Steel

3.2.1 Multilinear Isotropic Hardening (MISO)

The multilinear isotropic hardening is a very common isotropic hardening, which is characterized by a multilinear stress versus total or plastic strain curve (see Figure 3.3). It is defined in ANSYS by the following commands [1]:

Plastic Material Models of Metals

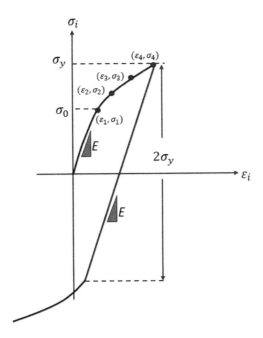

FIGURE 3.3
Stress versus strain for multilinear isotropic hardening.

```
TB, Plastic, matid,,,,MISO
TBPT, DEFI, 0, σ0
TBPT, DEFI, ε₁ᵖˡ, σ1
TBPT, DEFI, ε₂ᵖˡ, σ2
...
TBPT, DEFI, εᵢᵖˡ, σi
```

where ε_1^{pl}, ε_2^{pl}, ..., ε_i^{pl} appear in the TBPT commands.

One special form of multilinear isotropic hardening is the bilinear isotropic hardening model, which is a bilinear curve (see Figure 3.4). The definition of the bilinear hardening model in ANSYS is given as follows [1]:

```
TB, BISO, matid
TBDATA, 1, σ0 (Yield stress), E_T (Tangent modulus)
```

Here are the experimental data of 304 stainless steel [2] (see Figure 3.5), which are used to determine the material parameters of the bilinear isotropic hardening model as follows.

First, the Young's modulus E should be determined from the first four experimental data using the least square method:

$$E = 2169.2 \text{ MPa} \tag{3.4}$$

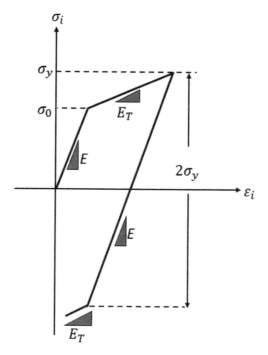

FIGURE 3.4
Stress versus strain for bilinear isotropic hardening.

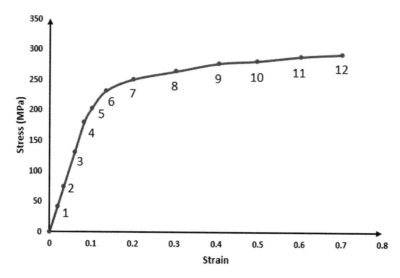

FIGURE 3.5
Experimental data of 304 stainless steel.

Plastic Material Models of Metals

Thus, the plastic strains are computed by

$$\varepsilon^{pl} = \varepsilon - \sigma/E \tag{3.5}$$

The curve of the plastic strains versus the stresses is plotted in Figure 3.6. These plastic data can be approximated by linear equation

$$y = 135.54x + 228.29 \tag{3.6}$$

Therefore, the yield stress $\sigma_0 = 228.29$ MPa, and $E_p = 135.54$ MPa.

The bilinear isotropic hardening model requires the elastoplastic tangent E_T, which is the slope of the total stress versus total strain. It can be converted from E_p by

$$E_T = \frac{E_p E}{E_p + E} = 127.57 \text{ MPa} \tag{3.7}$$

Thus, the bilinear isotropic hardening model of 304 stainless steel can be defined as

```
TB, BISO,1
TBDATA,1, 228.29, 127.57
```

Using the above material definition, the uniaxial tension was simulated in ANSYS. The results of the bilinear isotropic hardening model compared to

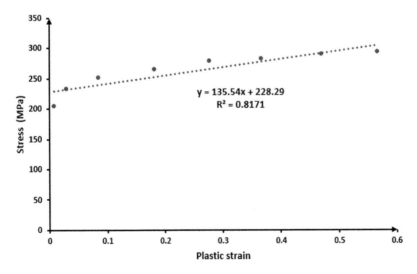

FIGURE 3.6
Plastic strain versus stress of 304 stainless steel.

those of the experimental data are illustrated in Figure 3.7, indicating that the bilinear isotropic hardening model greatly differs from the experimental data in the plastic stage. As a result, the bilinear isotropic hardening model is over-simplified for 304 stainless steel.

3.2.2 Voce Law Nonlinear Isotropic Hardening

Similar to bilinear isotropic hardening, the Voce hardening model adds an exponential saturation hardening term to the linear term (see Figure 3.8).

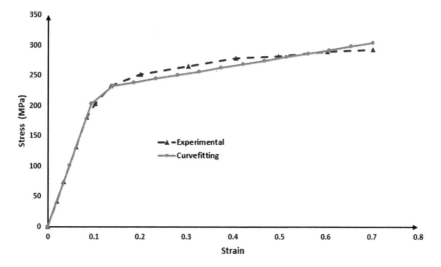

FIGURE 3.7
Curve-fitting results of bilinear isotropic hardening model.

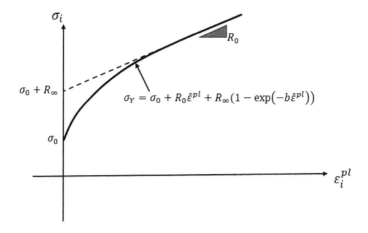

FIGURE 3.8
Stress versus plastic strain for Voce hardening model.

Plastic Material Models of Metals

The evolution of the yield stress in the Voce law nonlinear isotropic hardening has the following form [1]:

$$\sigma_Y = \sigma_0 + R_0 \hat{\varepsilon}^{pl} + R_\infty \left(1 - \exp\left(-b\hat{\varepsilon}^{pl}\right)\right) \quad (3.8)$$

The Voce law nonlinear isotropic hardening in ANSYS is specified as

```
TB, NLISO
TBDATA, 1, σ0, R0, R∞, b
```

Next, the 304 stainless steel was modeled using the Voce law nonlinear isotropic hardening. The Young's modulus is defined from the experimental data in the elastic stage as

$$E = 2169.2 \text{ MPa} \quad (3.9)$$

The yield stress is determined from the experimental points 5–8 using the second order polynomial (see Figure 3.9)

$$\sigma_0 = 201.59 \text{ MPa} \quad (3.10)$$

The linear approximation of the experimental points 9–12 has the following form (see Figure 3.10):

$$y = 55.17x + 262.55 \quad (3.11)$$

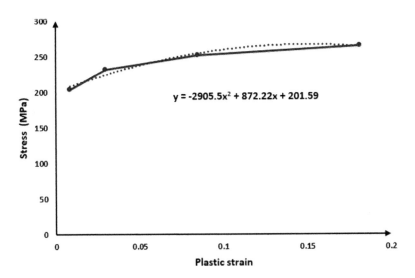

FIGURE 3.9
Curve-fitting for σ_0.

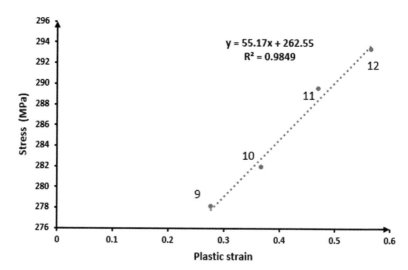

FIGURE 3.10
Linear approximation of the experimental points 9–12.

Therefore,

$$\sigma_0 + R_\infty = 262.55 \tag{3.12}$$

$$R_0 = 55.17 \tag{3.13}$$

From equation (3.10),

$$R_\infty = 262.55 - 201.59 = 61.0 \tag{3.14}$$

Value b is obtained from the experimental points 5–7

$$b = 16.44 \tag{3.15}$$

Thus, the Voce law nonlinear isotropic hardening is defined as

```
TB, NLISO
TBDATA, 1, 201.59, 55.17, 61.0, 16.44
```

The uniaxial tension was conducted in ANSYS using the above definition of the Voce law nonlinear isotropic hardening. The computational results against the experimental data are presented in Figure 3.11, showing that the computational results match the experimental data well. That demonstrates that 304 stainless steel can be modeled by the Voce law nonlinear isotropic hardening model.

Plastic Material Models of Metals

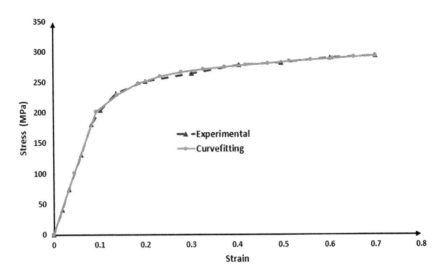

FIGURE 3.11
Curve-fitting results of Voce hardening model.

3.3 Nonlinear Kinematic Hardening

The Chaboche model has been widely used in industry to describe cyclic metal material behavior. Chaboche and Rousselier [3] found that the hardening behavior of the steel material can be better approximated by the sum of various Frederick–Armstrong formulas. The back-stress tensor in the Chaboche model is expressed as

$$a = \sum_{i=1}^{n} a_i \qquad (3.16)$$

where n is the number of superposed kinematic models.
The evolution of each back-stress model is specified as

$$\dot{a}_i = \frac{2}{3} C_i \dot{\varepsilon}^{pl} - \gamma_i \dot{\bar{\varepsilon}}^{pl} a_i + \frac{1}{C_i} \frac{dC_i}{d\theta} \dot{\theta} a_i \qquad (3.17)$$

where
  $\dot{\varepsilon}^{pl}$ – plastic strain rate,
  $\dot{\bar{\varepsilon}}^{pl}$ – magnitude of the plastic strain rate, and
  C_i and γ_i – user-input material parameters.

The Chaboche model is defined in ANSYS using commands

```
TB, Chaboche, 1
TBDATA, 1, σ0
TBDATA, 2, C1,γ1
TBDATA, 4, C2,γ2
...
TBDATA, 2n, Cn,γn
```

The Chaboche model can work with any of the available isotropic hardening models. It is very useful to simulate cyclic plastic behavior such as ratcheting.

Next, curve-fitting was conducted on the experimental data of one cycle using the curve-fitting tool available in ANSYS. The experimental data come from a low cycle fatigue test [4] (see Figure 3.12). First, the Young's modulus was determined from the elastic stage. Then, the plastic strain was obtained by subtracting the elastic strain from the total strain. With the available data of the plastic strain versus the stress, the curve-fitting was conducted in ANSYS using the following commands:

```
tbft,eadd,1,unia,ambi.exp
tbft,fadd,1,plas,chab,3          ! three-term Chaboche model
tbft,set,1,plas,chab,3,1,1e6     !initialize the parameters, C1
tbft,set,1,plas,chab,3,2,1e5     !initialize the parameters, γ1
...
tbft,set,1,plas,chab,3,7,55      !initialize σ0 = half of σmax
tbft,solve,1,plas,chab,3,0,1000
```

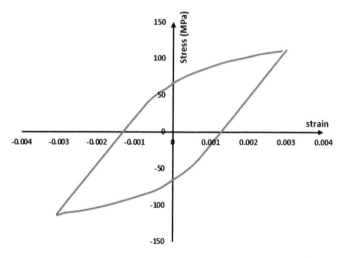

FIGURE 3.12
Experimental data for curve-fitting.

Plastic Material Models of Metals

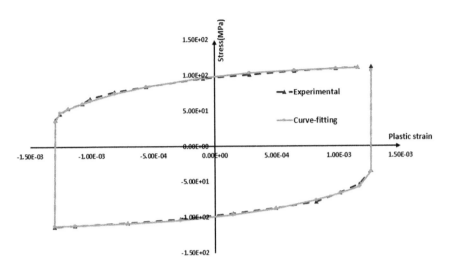

FIGURE 3.13
Curve-fitting results of Chaboche model.

In the above initial value input, σ_0 was defined from half of σ_{max}. Other initial data just chose big values.

Curve-fitting results, which are plotted in Figure 3.13, show that the curve-fitting results match the experimental data well. Therefore, the Chaboche model is defined as

```
TB, Chaboche, 1, ,3
tbdata, 1, 67.5
tbdata, 2, 1.000e+06, 9.371e+04
tbdata, 4, 1.077e+04, 9.994e+04
tbdata, 6, 4.110e+04, 1.111e+03
```

3.4 Summary

Chapter 3 introduces some isotropic hardening models and kinematic hardening models; it also provides examples to define the material parameters of these models from the curve-fitting of the experimental data.

References

1. ANSYS190 Help Documentation in the help page of product ANSYS190.
2. Wang, J.D., Ohno, N., Two equivalent forms of nonlinear kinematic hardening: application to nonisothermal plasticity, *International Journal of Plasticity*, Vol. 7, 1991, pp. 637–650.

3. Chaboche, J.L., Rousselier, G., On the plastic and viscoplastic constitutive equations, Parts I and II, *Journal of Pressure Vessel and Piping*, Vol. 105, 1983, pp. 153–164.
4. Seisenbacher, B., Winter, G., Grun, F., Improved approach to determine the material parameters for a combined hardening model, *Material Sciences and Applications*, Vol. 9, 2018, pp. 357–367.

4

Simulation of Metal Forming

Normally, when plasticity of material occurs, the material loses its strength. Therefore, most of the materials in practice are designed in the elastic range using the safety factor. On the other hand, material plasticity has been widely used in metal forming. Chapter 4 introduces metal forming and includes one example of its application.

4.1 Introduction of Metal Forming

Metal forming refers to the metalworking process of metal parts through mechanical deformation; although the shape of the workpiece changes, its mass remains the same [1]. The tool called a die applies stresses over the yield strength of the metal. Thus, the metal takes a shape determined by the shape of the dies (see Figure 4.1).

Generally, metal forming processes are classified into two groups: (1) bulk forming and (2) sheet-metal forming. Bulk forming processes specify that the work parts with low surface-area-to- volume ratios undergo significant deformations and shape changes. Bulk forming processes include rolling, forging, extrusion (see Figure 4.2), and drawing. The applied force in bulk deformation processing could be compressive, tensile, shear, or a combination of these forces. Unlike bulk forming, sheet metal forming has a high surface-area-to-volume ratio at the beginning. The sheet metal forming processes are accomplished in presses using a die or punch. Bending, drawing, and shearing (see Figure 4.3) constitute the major sheet metal forming operations.

The materials of metal forming require low yield strength and high ductility. In the metal forming process, the material is plastically deformed. Thus, it is that primary interest of the plastic region of the stress–strain curve of the metal in which the stress is a function of the exponent of plastic strain [2]

$$\sigma = K\varepsilon^n \qquad (4.1)$$

where K is strength coefficient, and n is strain hardening exponent.

These properties are strongly influenced by temperature. With the rise of the work temperature, the ductility of the metal increases but its yield strength decreases.

Next, the forming of a sheet between two dies was simulated in ANSYS190.

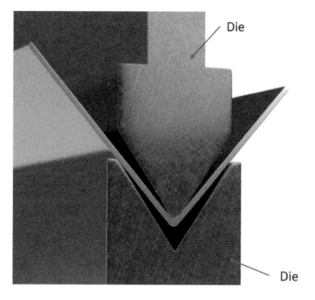

FIGURE 4.1
Forming of a sheet (Eugene Zabugin © 123rf.com).

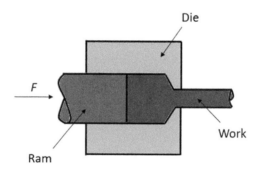

FIGURE 4.2
Basic bulk deformation processes of extrusion [2].

4.2 Simulation of Forming of a Sheet

4.2.1 Finite Element Model

A circle sheet was deformed between two dies. The right die was fixed, and the left one was moved down to make the sheet deformed. The whole model was simplified as a 2-D axisymmetrical model, and the sheet was meshed with Plane182 with keyopt(3)=1 (axisymmetrical) (see Figure 4.4).

Simulation of Metal Forming

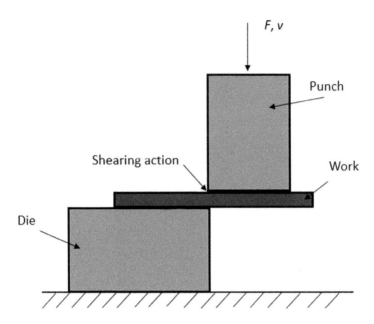

FIGURE 4.3
Basic sheet metalworking operation of shearing [2].

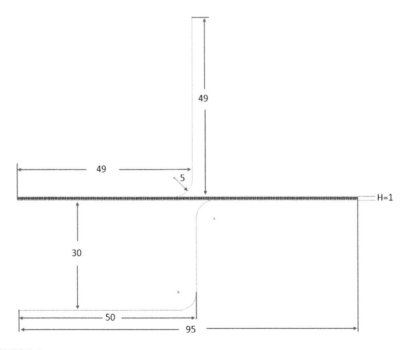

FIGURE 4.4
Finite element model of a sheet between two dies (all dimensions in mm).

TABLE 4.1

Material parameters of the sheet [3]

E(MPa)	v	σ_0 (MPa)	C_1	γ_1	R_0	R_∞	b	m_1	K_1
149,650	0.33	153	223	50	0	−153	317	1/7.7	1150

4.2.2 Material Properties

The sheet was modeled with the first-term Chaboche model plus the Exponential Visco-Hardening (EVH) option (see Table 4.1). The Chaboche model was defined in ANSYS by the following commands:

```
TB,CHAB,1,,1          ! DEFINE CHABOCHE MATERIAL DATA
TBDATA,1,1.53e2,62511,1.1/311*62511
```

The EVH Option was specified in ANSYS using commands

```
TB, RATE, 1,,6,EVH    ! DEFINE RATE DEPENDENT MATERIAL
                        DATA
TBDATA,1,1.53e2,0,-1.53e2,317,1/7.7,1150
```

4.2.3 Contact Definition

The contact between the sheet and the dies was assumed as a standard contact. The dies were regarded as rigid.

4.2.4 Loadings and Boundary Conditions

The right die was fixed with all DOFs, and the left one was moved down 30 mm (see Figure 4.5).

4.2.5 Solution Setting

In the solution setting, RATE should be turned on for the EVH option. Furthermore, the command NLGEOM ON and the maximum substeps 10,000 were defined because of the large deformation in the forming process. The entire forming process was assumed to be quasi-static without dynamic effect.

Simulation of Metal Forming

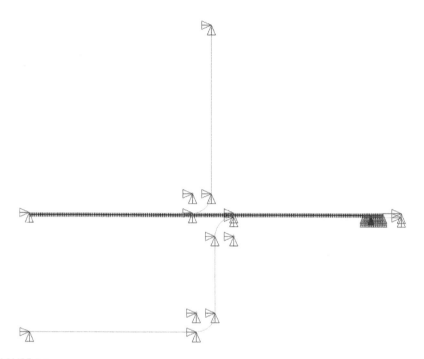

FIGURE 4.5
Boundary conditions.

4.2.6 Results

The deformation of the sheet, including its 3-D deformation, is plotted in Figure 4.6. von Mises stresses and plastic strains of the sheet at the last step are presented in Figures 4.7 and 4.8, respectively. The maximum stress and plastic strain are localized around the top corner with peak values of 1470 MPa and 0.412, respectively. Figure 4.9 illustrates the time history of the reaction force in the vertical direction, which indicates that the reaction force increases quickly at the beginning and reaches its peak value at time 0.333. After that, the reaction force reduces linearly.

4.2.7 Summary

The forming process of a sheet was modeled in ANSYS using the Chaboche model with the EVH option. The solution was complete, and the final stress and plastic strain state of the sheet were presented.

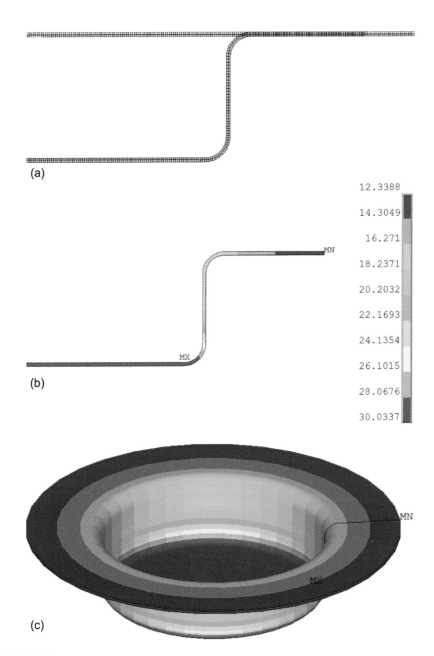

FIGURE 4.6
Final deformation of the sheet (a) deformed shape versus original shape (b) final displacement contour (2D axisymmetrical) (c) final displacement contour (3-D).

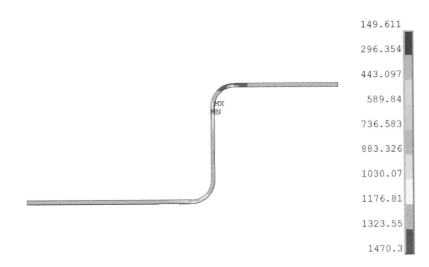

FIGURE 4.7
Final von Mises stresses of the sheet (MPa).

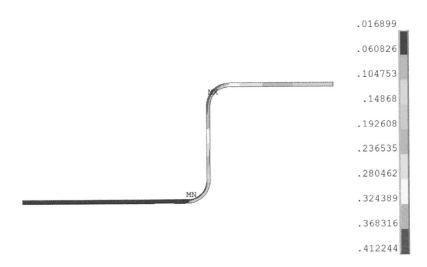

FIGURE 4.8
Final plastic strains of the sheet.

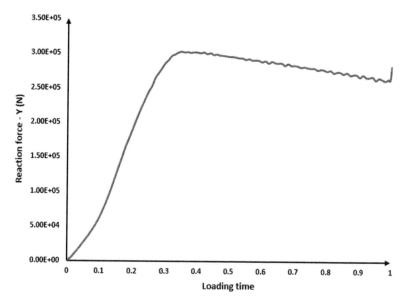

FIGURE 4.9
Time history of the reaction force in Y direction.

References

1. Lange, K., *Handbook of Metal Forming*, McGraw-Hill, Inc, 1985.
2. Groover, M.P., *Fundamentals of Modern Manufacturing*, 4th Edition, John Wiley & Sons, Inc, 2010.
3. Lin, R.C., Betten, J., and Brocks, W., Modeling of finite strain viscoplasticity based on the logarithmic corotational description, *Archive of Applied Mechanics*, Vol. 75, 2006, pp. 693–708.

5
Simulation of Ratcheting

Ratcheting is one unique feature of metal under cyclic loading. After its introduction, Chapter 5 then simulates ratcheting in a notched bar.

5.1 Introduction of Ratcheting

Ratcheting refers to a mechanical behavior that plastic deformation accumulates due to cyclic mechanical loading, which is characterized by a shift of the stress–strain hysteresis loop along the strain axis (see Figure 5.1) [1]. Thus, the combination of a steady state and cyclic loadings may produce ratcheting.

Figure 5.1 indicates that the strain increases with the cyclic number of the loading. Therefore, ratcheting may cause structural failure due to structural instability; because of this result, the study of ratcheting attracts more and more interest. The simulation of ratcheting demands that incremental plastic deformation terms are added to the general cyclic stress–strain equations. Thus, models like the multilinear kinematic hardening model cannot predict ratcheting. A few models are proposed to simulate ratcheting; the Chaboche model discussed in Section 3.3 is the most widely used and appears in the commercial software ANSYS [2]. Next, an example of ratcheting simulation using the Chaboche model in ANSYS190 is presented.

5.2 Simulation of Ratcheting in a Notched Bar

In this example, a notched bar subjected to cycling loading was simulated to produce the plastic ratcheting in ANSYS190.

5.2.1 Finite Element Model

A 2-D axisymmetrical notched rod was modeled in ANSYS190 using Plane182 with keyopt (3) = 1 (axisymmetrical) (see Figure 5.2). Also, the model was simplified using the symmetrical condition in the vertical direction.

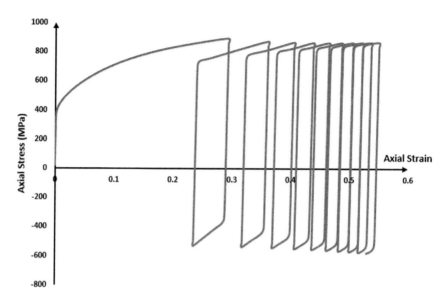

FIGURE 5.1
Ratcheting.

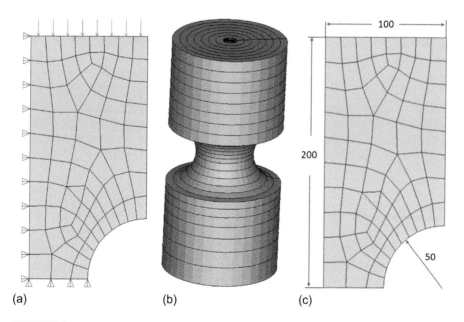

FIGURE 5.2
Finite element model of a notched rod (a) 2-D axisymmetrical (b) 3-D (c) Geometry (all dimensions in mm).

Simulation of Ratcheting

TABLE 5.1

Material parameters of the rod [3]

E(MPa)	v	σ_0 (MPa)	C_1	γ_1	C_2	γ_2	C_3	γ_3	n
647,000	0.3	67.5	1e6	9.37e4	1e4	1e5	4.1e4	1.1e3	0.1

5.2.2 Material Properties

The rod was defined with the three-term Chaboche model plus the Voce isotropic hardening law (see Table 5.1). The material parameters of the Chaboche model were selected from the curve-fitting results of Section 3.3.

The material model was defined by the following commands:

```
POWER_N=0.1
SIGMA_Y=67.5
TB,NLISO,1,,2,5
TBDATA,1,SIGMA_Y,POWER_N
tb,chab,1,,3
tbdata,1,SIGMA_Y,1e6,9.37e4,1e4,1e5
tbdata,6,4.1e4,1.1e3
```

For comparison, the model was repeated with the bilinear kinematic hardening model, in which the tangent modulus E_T was chosen for 2000MPa.

5.2.3 Loadings and Boundary Conditions

The axisymmetrical condition and symmetrical condition were applied on edges (see Figure 5.2a). The top edge was loaded with ten cycles of loadings; each cycle had a maximum of 50MPa in tension and a minimum of 30MPa in compression (see Figure 5.3).

5.2.4 Results

The von Mises plastic strain and stress of the rod at the end of the first cycle and the last cycle are plotted in Figures 5.4 and 5.5, respectively. The maximum values occur at the notched tip for both stresses and plastic strains. Comparing the results of the first cycle and last cycle, the stresses do not change much because they are pressure loading, but the plastic strains change dramatically between the first cycle and the last cycle. Figures 5.6 and 5.7 depict the axial strain versus axial stress, with the Chaboche model and bilinear kinematic hardening model, respectively. Figure 5.6 clearly illustrates a shift of the stress–strain hysteresis loop along the strain axis, which is the ratcheting. On the other hand, no ratcheting appears in Figure 5.7. The difference between Figures 5.6 and 5.7 confirms that the ratcheting is due to the Chaboche model (nonlinear kinematic hardening), and the bilinear kinematic hardening model cannot produce the ratcheting of metal.

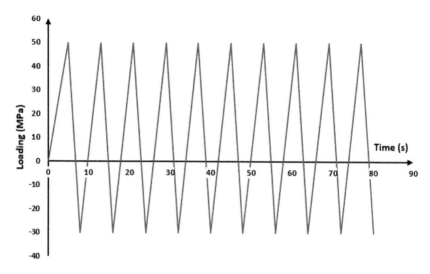

FIGURE 5.3
Loading history.

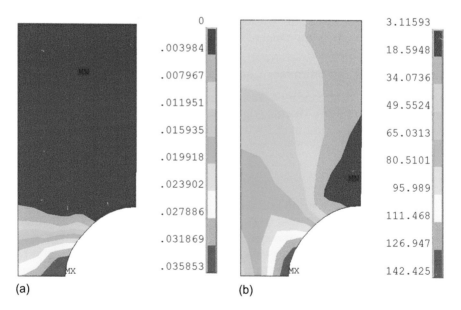

FIGURE 5.4
vM plastic strains and stresses of the rod at the end of 1st cycle (a) vM plastic strains (b) vM stresses (MPa).

Simulation of Ratcheting

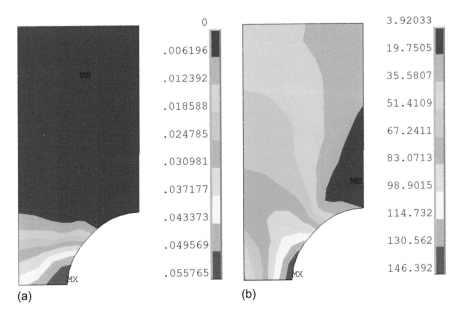

FIGURE 5.5
vM plastic strains and stresses of the rod at the end of the last cycle (a) vM plastic strains (b) vM stresses (MPa).

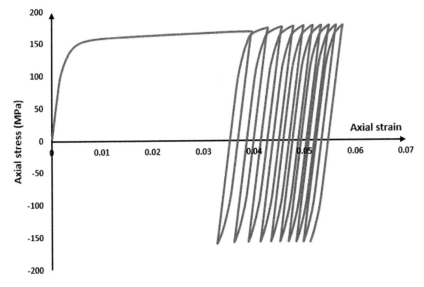

FIGURE 5.6
Axial strain versus axial stress.

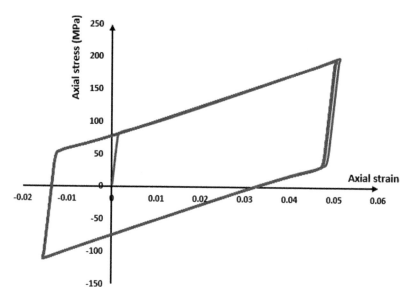

FIGURE 5.7
Axial strain versus axial stress with bilinear kinematic hardening.

5.2.5 Summary

A notched bar with the Chaboche model under cycling loading was studied in ANSYS190. The relation between the axial strain and axial stress reveals the ratcheting due to the Chaboche model (nonlinear kinematic hardening).

References

1. Bree, J., Elastic-plastic behavior of thin tubes subjected to internal pressure and intermittent high-heat fluxes with application to fast-nuclear-reactor fuel elements, *Journal of Strain Analysis*, Vol. 2, 1967, pp. 226–238.
2. ANSYS190 Help Documentation in the help page of product ANSYS190.
3. Kang, G., Finite element implementation of advanced constitutive model emphasizing on ratchetting, 18th International Conference on Structural Mechanics in Reactor Technology (SMiRT 18) Beijing, China, August 7–12, 2005.

6

Influence of Temperature on Material Properties

Chapter 6 focuses on the relationship between temperature and material properties, including the definition of temperature dependency of material properties and simulation of a combustion chamber.

6.1 Temperature Dependency of Material Properties

Material properties of metals/alloys are strongly temperature-dependent [1]. By lowering the temperature, the number and tightness of chemical bonds increase. Thus, yielding becomes more and more difficult, and the initial yield stress σ_0 becomes a larger value. Other plastic parameters have similar change with temperature. In ANSYS, the temperature dependency is defined using command TBTEMP. Here is one example of the Chaboche model with different temperatures:

```
tb,chab,1,n,,
tbtemp,T1
tbdata,1,σ₀ (T1),C₁ (T1),γ₁ (T1),···
tbtemp,T2
tbdata,1,σ₀ (T2),C₁ (T2),γ₁ (T2),···
...
tbtemp,Tn
tbdata,1,σ₀ (Tn),C₁ (Tn),γ₁ (Tn),···
```

Temperature affects the mechanical behavior of metals/alloys in many aspects. The temperature directly causes thermal strain. Also, as mentioned in the above paragraph, the material properties of the metals/alloys are functions of temperature. Normally, with the increase of the temperature, the Young's modulus and yield stress decrease. In addition, the governing equations of the material model such as equation (3.17) for the back stress of the Chaboche model have terms to represent the influence of temperature change. Section 6.2 simulates the combustion chamber to illustrate these effects.

6.2 Simulation of Combustion Chamber under Different Temperatures

Here, the combustion chamber was simulated in ANSYS190 to study its mechanical behavior with temperature loading.

6.2.1 Finite Element Model

A quarter of the combustion chamber was modeled using Plane182 with kyopt(3)=2 (plane strain condition) (see Figure 6.1).

6.2.2 Material Properties

The material of the combustion chamber was defined with the one-term Chaboche model plus the Perzyna option. Also, the material properties of the combustion chamber vary with temperatures (Table 6.1) [2]. These material models were defined in ANSYS as follows:

```
! define Young's modulus and Poisson's ratio
tb,elas,1,3
tbtemp,20
tbdata,1,76000,0.33
tbtemp,100
tbdata,1,72000,0.33
```

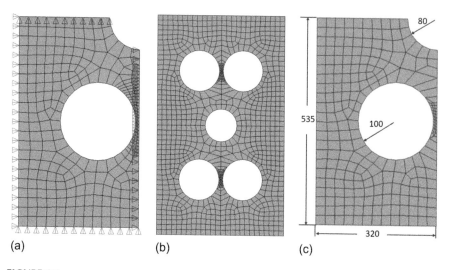

FIGURE 6.1
Finite element model of a combustion chamber. (a) A quarter of the model (b) Full model (c) Geometry (all dimensions in mm).

TABLE 6.1

Material parameters of the combustion chamber [2]

Temperature (degree)	20	100	200	300
E(MPa)	76,000	72,000		60,000
Chaboche Model				
σ_0(MPa)	80	63	52	4
C_1	21,700	22,000	8,460	3,780
γ_1	155	200	310	420
Perzyna				
m_1	1/8.5		1/8.5	1/8.5
γ_1	1e-4		3e-6	1e-10
MISO				
σ_0(MPa)	80	63	52	4
$\sigma(\varepsilon=0.1)$ (MPa)	200	157	130	10

```
tbtemp,300
tbdata,1,60000,0.33
! define Chaboche model
tb,chab,1,4,1
tbtemp,20
tbdata,1,80,21700,155
tbtemp,100
tbdata,1,63,22000,200
tbtemp,200
tbdata,1,52,8460,310
tbtemp,300
tbdata,1,4,3780,420
! define Perzyna model
TB,RATE,1,3,,PERZYNA        ! RATE Table
tbtemp,20
TBDATA,1,1/8.5,1e-4
tbtemp,200
tbdata,1,1/8.5,3e-6
tbtemp,300
tbdata,1,1/8.5,1e-10
mp,alpx,1,6e-6
! define MISO
tb,plas,1,4,,MISO
TBTEMP,20
TBPT,DEFI,0,80
TBPT,DEFI,0.1,200
TBTEMP,100
TBPT,DEFI,0,63
```

```
TBPT,DEFI,0.1,157
TBTEMP,200
TBPT,DEFI,0,52
TBPT,DEFI,0.1,130
TBTEMP,300
TBPT,DEFI,0,4
TBPT,DEFI,0.1,10
```

6.2.3 Loadings and Boundary Conditions

The combustion chamber was fixed with all sides. In addition, symmetrical conditions were applied on the edges (see Figure 6.1a). The temperature of the combustion chamber changes with time (see Figure 6.2).

6.2.4 Results

The von Mises plastic strain and stress of the combustion chamber at each loading step are plotted in Figures 6.3 through 6.8, respectively. The maximum values occur at the hole edge with the narrowest width and vary at different loading steps. At that location, the time history of the stress, plastic strain, and the thermal strain are plotted in Figure 6.9.

6.2.5 Discussion

A combustion chamber was simulated under different temperatures with material properties as a function of temperature. Obviously, temperature

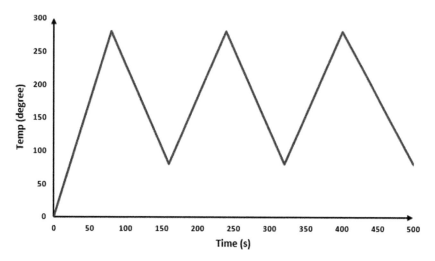

FIGURE 6.2
Loading history of temperature.

Influence of Temperature on Material Properties

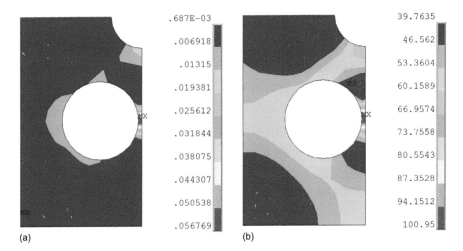

FIGURE 6.3
vM plastic strains and stresses of the combustion chamber at time 80s. (a) vM plastic strains (b) vM stresses (MPa).

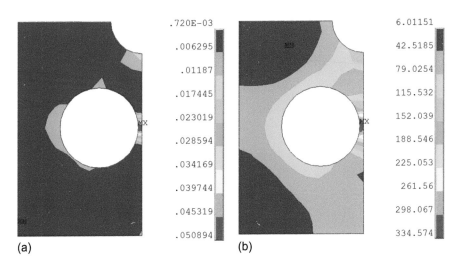

FIGURE 6.4
vM plastic strains and stresses of the combustion chamber at time 160s. (a) vM plastic strains (b) vM stresses (MPa).

plays a major role in the model. The temperature causes the thermal strain, which directly affects the stresses and strains of the combustion chamber. At the first loading step, the thermal strain caused by the temperature change makes the stress rise quickly. After reaching yield (point a in Figure 6.9c), the plastic strain increases as well. Because the high temperature corresponds to

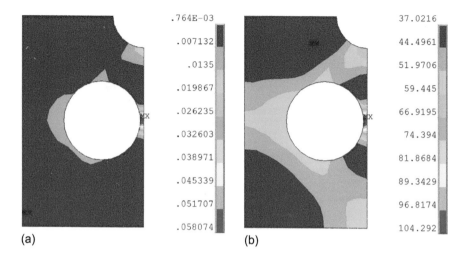

FIGURE 6.5
vM plastic strains and stresses of the combustion chamber at time 240s. (a) vM plastic strains (b) vM stresses (MPa).

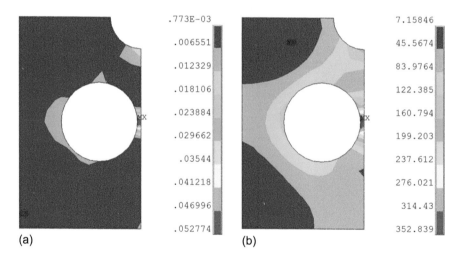

FIGURE 6.6
vM plastic strains and stresses of the combustion chamber at time 320s. (a) vM plastic strains (b) vM stresses (MPa).

a low yield stress, the stress drops at the later stage of the first loading step, although the plastic strain keeps rising. After the temperature decreases in the second loading step, the material is in the elastic stage between points b and c in Figure 6.9c. After reaching the yield surface, the plastic strain changes between points c and d in Figure 6.9c until the end of the second

Influence of Temperature on Material Properties

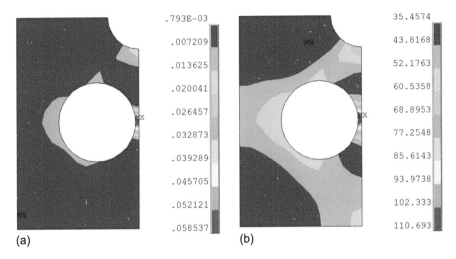

FIGURE 6.7
vM plastic strains and stresses of the combustion chamber at time 400s. (a) vM plastic strains (b) vM stresses (MPa).

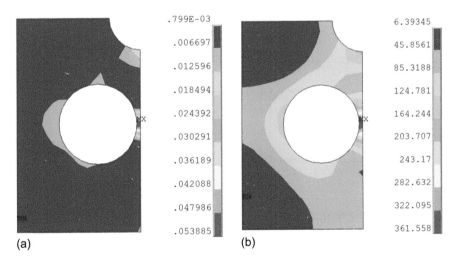

FIGURE 6.8
vM plastic strains and stresses of the combustion chamber at time 500s. (a) vM plastic strains (b) vM stresses (MPa).

loading step. The similar behavior of the stresses and strains of the combustion chamber occurs at the later loading steps. It is also observed that at the beginning from o to point a in Figure 6.9c, the material is in the elastic stage. With the rise of temperature, the Young's modulus decreases, which explains that the stress in this period increases nonlinearly (see Figure 6.9b)

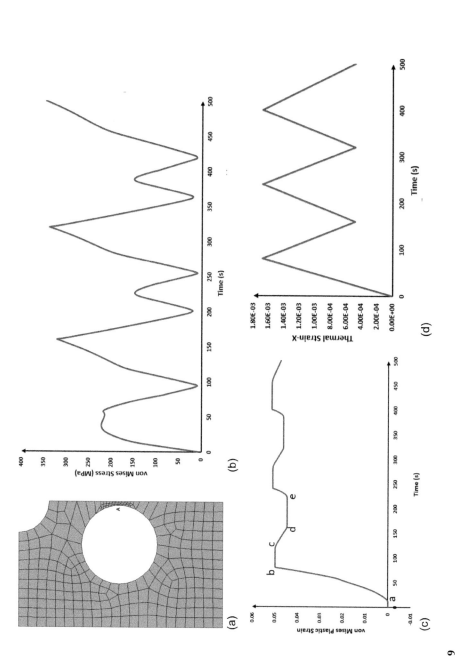

FIGURE 6.9
Results of element A with time (a) element A (b) vM stress of element A with time (c) vM plastic strain of element A with time (d) thermal strain of element A with time.

Influence of Temperature on Material Properties

although the material is in the elastic stage. Furthermore, some terms in equation of the Chaboche model (3.17) are associated with temperature. Overall, the temperature plays three roles in the model: 1) thermal strain, 2) material properties as a function of temperature, and 3) thermal term in the Chaboche equation.

6.2.6 Summary

A combustion chamber was simulated under different temperatures. The computational results indicate that temperature affects the mechanical behavior of the combustion chamber in three aspects: (1) thermal strain, (2) material properties as a function of temperature, and (3) thermal term in the Chaboche equation.

References

1. Roylance, D., *Mechanical Properties of Materials*, MIT, 2008.
2. http://z-mat.com/PDFs/info/z-mat-detail.pdf.

7
Simulation of Creep

Creep exists in metal, especially under high temperature. In Chapter 7, creep is modeled by both ANSYS and user subroutine.

7.1 Introduction of Creep

7.1.1 Creep

Creep refers to the strains (stresses) of metals that, when under high temperatures, change with time in constant working conditions. Figure 7.1 illustrates the typical creep curve composed of three stages [1]. The first stage is the primary creep stage in which the creep starts at a rapid rate and slows gradually with time. The creep maintains a relatively uniform rate in the second creep stage; thus, the second creep stage is the steady stage. The third stage has an accelerating creep ratio that always leads to the material's failure. Creep always occurs in metal working, springs, soldered joints, and high-temperature materials. Two major cases exist relevant to creep:

(1) Under the constant loading, the creep strains of the structure increase with time. Thus, the deformations of the structure are getting larger and larger (see Figure 7.2a).
(2) Under initial stresses and fixed boundary conditions, stresses of the structure decrease with time. That is called stress relaxation (see Figure 7.2b).

7.1.2 Creep Constitutive Law

Generally, the creep equation is a function of stress, strain, time, and temperature,

$$\dot{\varepsilon}_{cr} = f_1(\sigma)f_2(\varepsilon)f_3(t)f_4(T) \tag{7.1}$$

ANSYS provides 13 implicit creep equations like time hardening. The time hardening equation is expressed as [2]

$$\dot{\varepsilon}_{cr} = C_1 \sigma^{c2} t^{c3} e^{-c4/T} \tag{7.2}$$

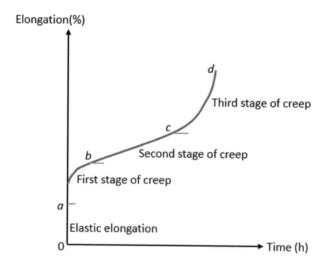

FIGURE 7.1
Typical creep curve.

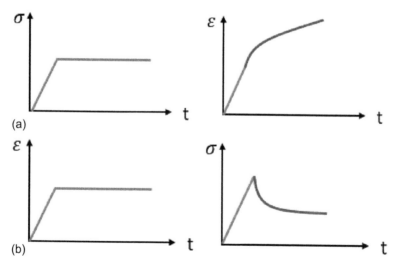

FIGURE 7.2
Two major creep behaviors (a) creep (b) stress relaxation.

The equation (7.2) indicates that the time hardening equation has four parameters – C1, C2, C3, and C4 – related to stress, time, and temperature, but no strain. C2 defines the relation between the creep ratio and stress. C2 is normally selected between 2 and 20. C3 is the time factor. Negative C3 means a decrease of creep ratio with time, and vice versa; thus, C3 in the first creep stage should be negative.

Simulation of Creep 49

The implicit creep analysis in ANSYS is robust, fast, accurate, and widely used in practice. The creep function can be used individually or combined with kinematic hardening plasticity. The ANSYS command to define the creep function is specified as follows [2]:

```
TB, CREEP, matid,,,TBOPT   ! TBOPT specifies the creep
equation
TBDATA, 1, C1, C2, C3, C4
```

7.1.3 Subroutine UserCreep

In the creep option of TB command, if TBOPT=100, it is UserCreep. ANSYS provides this option for customers to develop their own creep function. The UserCreep subroutine demands the output of incremental creep strain as well as the derivative of incremental creep strain with respect to effective stress and effective creep strain, respectively [2]. The time hardening equation (7.2) was developed as an example of UserCreep. Its derivative of incremental creep strain with respect to effective stress and effective creep strain is computed by

$$\frac{d(\dot{\varepsilon}_{cr})}{d\sigma} = c2 \times C_1 \sigma^{c2-1} t^{c3} e^{-\frac{c4}{T}} = c2 \times \frac{\dot{\varepsilon}_{cr}}{\sigma} \tag{7.3}$$

$$\frac{d(\dot{\varepsilon}_{cr})}{d\varepsilon_{cr}} = 0 \tag{7.4}$$

Equations (7.2)–(7.4) are implemented in the following subroutine:

```
*deck,usercreep  USERDISTRIB parallel
   SUBROUTINE usercreep (impflg, ldstep, isubst, matId,
   elemId,
 &            kDInPt, kLayer, kSecPt, nstatv, nprop,
 &            prop , time , dtime , temp , dtemp ,
 &            toffst, Ustatev, creqv , pres , seqv ,
 &            delcr , dcrda)
c************************************************************
c  *** primary function ***
c         Define creep laws when creep table options are
c         TB,CREEP with TBOPT=100.
c         Demonstrate how to implement the UserCreep
c         subroutine
c
c         Creep equation is
c            dotcreq := k0 * seqv ^ n * time ^ m * exp (-b/T)
c
```

```
c           seqv is equivalent effective stress (Von-Mises
            stress)
c           creqv is equivalent effective creep strain
c           T    is the temperature
c           k0, m, n, b are material constants
c
c     This model corresponds to the time hardening creep
      function TBOPT = 2
c
c
c
c*************************************************************
c
c     input arguments
c     ===============
c     impflg   (in  ,sc   ,i)        Explicit/implicit
                                     integration
c                                    flag (currently not used)
c     ldstep   (in  ,sc   ,i)        Current load step
c     isubst   (in  ,sc   ,i)        Current substep
c     matId    (in  ,sc   ,i)        Number of material index
c     elemId   (in  ,sc   ,i)        Element number
c     kDInPt   (in  ,sc   ,i)        Material integration point
c     kLayer   (in  ,sc   ,i)        Layer number
c     kSecPt   (in  ,sc   ,i)        Section point
c     nstatv   (in  ,sc   ,i)        Number of state variables
c     nprop    (in  ,sc   ,i)        Size of mat properties
                                     array
c
c     prop     (dp  ,ar(*),i)        Mat properties array
c     time                           Current time
c     dtime                          Current time increment
c     temp                           Current temperature
c     dtemp                          Current temperature
                                     increment
c     toffst   (dp, sc, i)           Temperature offset from
                                     absolute zero
c     seqv     (dp ,sc , i)          Equivalent effective stress
c     creqv    (dp ,sc , i)          Equivalent effective creep
                                     strain
c     pres     (dp ,sc , i) Hydrostatic pressure stress,
                            -(Sxx+Syy+Szz)/3
c               Note that: Constitutive consistency is
                           not accounted for
```

```
c                     if creep strains are function of pressure
c
c    input output arguments        input desc    / output desc
c    ======================        ==========      ==========
c    Ustatev (dp,ar(*), i/o)    user defined internal
c                                   state variables at
c                          time 't' / 't+dt'.
c                       This array will be passed in containing the
c                       values of these variables at start of the
c                       time increment. They must be updated in this
c                       subroutine to their values at the end of
c                       time increment, if any of these internal
c                       state variables are associated with the
c                       creep behavior.
c
c    output arguments
c    ================
c    delcr   (dp ,sc , o)       incremental creep strain
c    dcrda   (dp,ar(*), o)        output array
c                               dcrda(1) - derivative of
c                                 incremental creep
c                                 strain to effective stress
c                               dcrda(2) - derivative of
c                                 incremental creep
c                                 strain to creep strain
c
c    local variables
c    ===============
c    c1,c2,c3,c4 (dp, sc, 1)    temporary variables as
c                                 creep constants
c    con1            (dp ,sc, 1)    temporary variable
c    t               (dp ,sc, 1)    temporary variable
c
c***********************************************************
c
c --- parameters
c
#include "impcom.inc"
.
      DOUBLE PRECISION ZERO
      PARAMETER    (ZERO = 0.0d0)
c
c --- argument list
c
```

```
      INTEGER             ldstep, isubst, matId , elemId,
     &                    kDInPt, kLayer, kSecPt, nstatv,
     &                    impflg, nprop
      DOUBLE PRECISION dtime , time , temp , dtemp ,
                          toffst,
     &                    creqv , seqv , pres
      DOUBLE PRECISION prop(*), dcrda(*), Ustatev(nstatv)
c
c --- local variables
c
      DOUBLE PRECISION c1   , c2   , c3   , c4   ,
     &                    con1 , delcr , t
c
c*************************************************************
c
c *** Skip when stress and creep strain are both zero
      if (seqv.LE.ZERO.AND.creqv.LE.ZERO) GO TO 990
c *** Add temperature off set
      t    = temp + toffst
c *** Primary creep function
c     delcr := c1 * seqv ^ n * time ^ m * exp
      (-b/T) * dtime
      c1   = prop(1)
      c2   = prop(2)
      c3   = prop(3)
      c4   = prop(4)
c *** User needs to make sure if c4 has nonzero value,
      then temperature
c should be also nonzero.
      con1 = ZERO
      if(c4.ne.ZERO .and. t.gt.ZERO) con1 = c4/t
c *** Calculate incremental creep strain
      if (creqv .le. TINY) creqv = sqrt(TINY)
      delcr = ZERO
      IF(c1.gt.ZERO) delcr  = (exp( log(c1) + c2 *
      log(seqv) +
     &                    c3 * log(time) - con1 )) * dtime
c *** Derivative of incremental creep strain to
      effective stress
      dcrda(1)= c2 * delcr / seqv
c *** Derivative of incremental creep strain to
      effective creep strain
      dcrda(2)= 0
```

Simulation of Creep

```
c *** Write the effective creep strain to last state
      variable
      if (nstatv .gt. 0) then
         Ustatev(nstatv) = creqv
      end if
 990  continue
      return
      end
```

7.2 Simulation of Creep of a Bolt under Pretension

In Section 7.2, the creep of a bolt under pretension was studied in ANSYS190 using TB, CREEP command, and UserCreep.

7.2.1 Finite Element Model

The model is two steel plates connected by a bolt. Because of the symmetrical condition, a half of the model was created and meshed by SOLID187 (see Figure 7.3).

7.2.2 Material Properties

Both the bolt and the steel plate were modeled as a MISO plastic model. In addition, the creep of the bolt was defined as time hardening using TB,

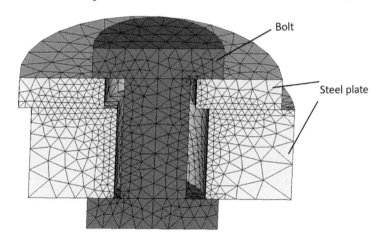

FIGURE 7.3
Finite element model of pretension between the plates and the bolt.

CREEP command, and UserCreep for comparison. The material models were defined using the following APDL commands:

```
! material model of steel
mp,ex,1,14.665e6
mp,prxy,1,0.30
TB,PLAS,1,1,2,MISO
tbPT,defi,0,29.33e3
TBPT,defi,1.59e-3,50e3

! material model of bolt
mp,ex,2,42e6
mp,prxy,2,0.30
TB,PLAS,2,1,2,MISO
tbPT,defi,0,29.33e3*3
TBPT,defi,1.59e-3,50e3*3
TB,CREEP,2,1,4,100
tbdata,1,0.54e-16,2.3,-0.6
```

7.2.3 Loadings and Boundary Conditions

In addition to the symmetrical condition, 0.01inch pretension was applied on the bolt using the SLOAD command (see Figure 7.4).

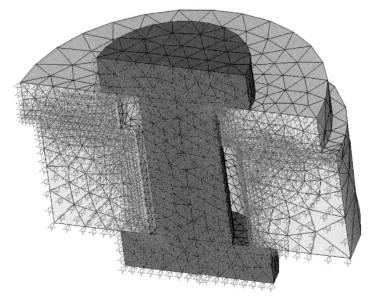

FIGURE 7.4
Boundary conditions.

Simulation of Creep 55

7.2.4 Solution Setting

The solution has two steps. The first step is as short as 1e−4 seconds to load the pretension. In the first step, RATE was turned off to deactivate the creep because the creep calculation is always inaccurate during a very short time period. In the second step, RATE was turned on, and the time in the second step is creep time.

7.2.5 Results

Command TB, CREEP and UserCreep got the identical results. Here, only the results obtained by TB, CREEP are presented.

Figures 7.5 and 7.6 illustrate the vM stresses of the model at the end of the first step and the second step, respectively. The results indicate that stresses at the end of the last step are much less than those at the end of the first step, which is due to the creep of the bolt. This conclusion is confirmed by the time history of von Mises stress of an element of the bolt (see Figure 7.7) that illustrates the case of stress relaxation mentioned in Figure 7.2b.

7.2.6 Discussion

Command TB, CREEP and UserCreep got the identical results, which validate the developed UserCreep. Moreover, the creep simulation is divided

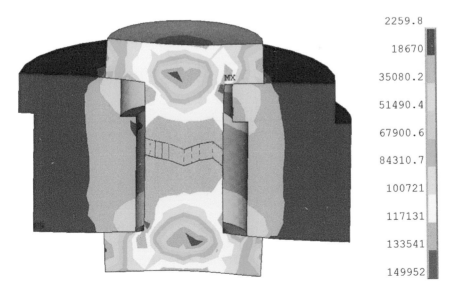

FIGURE 7.5
vM stresses of the model after pretension (1st loading step) (Psi).

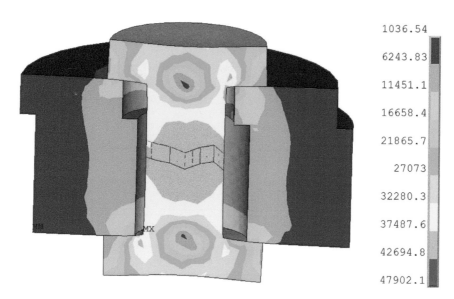

FIGURE 7.6
vM stresses of the model after 17.36 days (Psi).

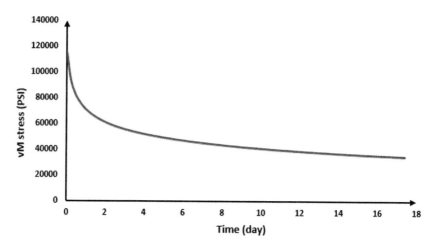

FIGURE 7.7
vM stress of one element of the bolt with time.

into two steps. The first step is the loading step without creep effect. Normally, the time of the first step is very short. The second step is the creep step with RATE turned on. The method of dividing the creep calculation into a loading step and creep step is widely used in the creep calculation because it is proven that the calculation of creep strain may be inaccurate in a very short time.

7.2.7 Summary

The creep of a bolt with pretension was simulated in ANSYS190. Command TB, CREEP and UserCreep have the same results that validate the developed UserCreep.

References

1. Meyers, M., Chawla, K., *Mechanical Behavior of Materials*, Cambridge University Press, 2008.
2. ANSYS190 Help Documentation in the help page of product ANSYS190.

Part II

Polymers

Polymers with a molecular structure exhibit large deformation, creep, plastic flow, viscoelastic flow, rupture, and wear. To capture the behaviors of polymers, some hyperelastic material models are available in ANSYS that are discussed in Part II.

Chapter 8 first presents the structure and features of polymers; Chapter 9 then focuses on some hyperelastic models and their application such as simulation of breast shape after breast surgery. Chapter 10 emphasizes the viscoelasticity of polymers, including simulation of the liver soft tissues. The elastomer damage Mullins effect is examined in Chapter 11. Chapter 12 exhibits a simple and useful way to model hyperelastic materials using subroutine UserHyper in ANSYS.

8

Structure and Features of Polymer

Chapter 8 briefly introduces the structure and features of polymer.

8.1 Structure of Polymer

A polymer is defined as a material consisting of many macromolecules. Each macromolecule is a long chain linked from a number of monomer units formed by covalent bonds. Thus, a polymer is formed with three different length scales. In the first scale, the most local scale, the atoms are arrayed into monomer units and joined together by covalent bonds. The other two scales are the long chain-like structure and network structure. The different chain molecules have three primary forms (see Figure 8.1) [1]: 1) linear macromolecule, 2) branched macromolecule, and 3) crosslinked macromolecule.

Different structures lead to various material properties. The linear macromolecule becomes soft when heated and hard when cooled, which is called thermoplasticity. On the other hand, the crosslinked macromolecule cannot flow or melt easily with heat because its relative motion among the molecular chains is limited; that is referred to as thermoset.

8.2 Features of Polymer

Compared with metal, polymer has the following features:

(1) The network structures interact by weak van der Waals forces, resulting in low stiffness and high ductility of polymer. The elastic modulus of polymer is about 1e6 times less than that of metal.
(2) The Poisson's ratio of polymer is around 0.49, close to 0.50, which means polymer has a very high bulk modulus.
(3) Uncrosslinked polymer reveals time-dependent elastic deformation, namely relaxation.
(4) Temperature influences all aspects of the mechanical behaviors of polymer. Figure 8.2 plots variations of the material's stiffness (Young's modulus) with temperature [2].

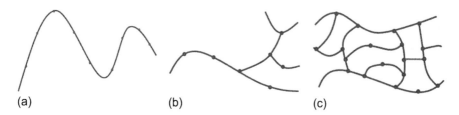

FIGURE 8.1
Structures of a single macromolecular chain. (a) Linear macromolecule (b) Branched macromolecule (c) Crosslinked macromolecule.

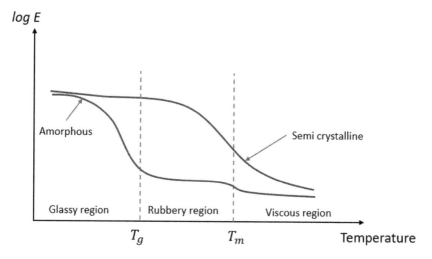

FIGURE 8.2
Variety of the Young's modulus with temperature for different classes of elastomers [2]. T_g – glass transition temperature; T_m – melting temperature.

References

1. Zhang, S., *Tribology of Elastomers*, Elsevier, 2004.
2. Bergstrom, J., *Mechanics of Solid Polymers*, Elsevier, 2015.

9
Hyperelasticity

After presenting some common hyperplastic models and conducting the curve-fitting of material parameters, Chapter 9 then simulates the rubber rod under compression and breast shape under gravity after breast surgery.

9.1 Some Widely Used Hyperelastic Models

9.1.1 Neo-Hookean Model

The Neo-Hookean model is a simple hyperelastic model with two material parameters: a shear modulus μ and a bulk modulus κ. The elastic energy density of the Neo-Hookean model is expressed as [1]

$$W = \frac{\mu}{2}(I_1 - 3) + \frac{1}{d}(J - 1)^2 \tag{9.1}$$

The equation is a linear function of I_1. Thus, it cannot catch the large-strain nonlinear behavior of many elastomers. In addition, no I_2 term in the energy equation makes stress prediction low when the biaxial loading dominates. Overall, the advantage of the Neo-Hookean model is that the model is robust and computationally efficient; its limitation is that it cannot give an accurate prediction in case of a large strain.

9.1.2 Mooney–Rivlin Model

Unlike the Neo-Hookean model, the Mooney–Rivlin model includes a linear term of I_2 and has the following form [1]:

$$W = c_{10}(I_1 - 3) + c_{01}(I_2 - 3) + \frac{1}{d}(J - 1)^2 \tag{9.2}$$

The Mooney–Rivlin model may improve the stress prediction, but it also raises another issue – having a negative c_{01} term, which may cause the model to be unstable in the finite deformation.

9.1.3 Yeoh Model

The Yeoh model surpasses the Neo-Hookean model because it has a high order of I_1 without I_2 [1]

$$W = \sum_{i=1}^{N} C_{i0}(I_1 - 3)^i + \sum_{k}^{N} \frac{1}{d_k}(J-1)^{2k} \tag{9.3}$$

Therefore, the Yeoh model can predict the stress more accurately than the Neo-Hookean model; it can also avoid the stability issue of the Mooney–Rivlin model.

9.1.4 Polynomial Model

The polynomial model – a generalization of the Neo-Hookean model, the Mooney–Rivlin model, and the Yeoh Model – involves taking a polynomial expansion in terms of I_1 and I_2 as [1]

$$W = \sum_{i+j=1}^{N} C_{ij}(I_1 - 3)^i (I_2 - 3)^j + \sum_{k}^{N} \frac{1}{d_k}(J-1)^{2k} \tag{9.4}$$

The disadvantage of the Polynomial model is that it is very difficult to determine the material parameters to make an accurate and robust computation in multiaxial loading cases.

9.1.5 Gent Model

The Gent model, an extension of the Neo-Hookean model, aims to capture the large-strain behavior of elastomer-like material expressed as [1]

$$W = -\frac{\mu J_m}{2} \ln\left(1 - \frac{I_1 - 3}{J_m}\right) + \frac{1}{d}\left(\frac{J^2 - 1}{2} - \ln J\right) \tag{9.5}$$

In the above equation, the Gent model has one extra parameter J_m to control the limited chain extensibility at large strains.

9.1.6 Ogden Model

The Ogden model differs from the other hyperelastic materials because the energy density of Ogden is expressed in terms of the applied principal stretches [1]

$$W = \sum_{i=1}^{N} \frac{\mu_i}{\alpha_i}\left(\bar{\lambda}_1^{\alpha_i} + \bar{\lambda}_2^{\alpha_i} + \bar{\lambda}_3^{\alpha_i} - 3\right) + \sum_{k}^{N} \frac{1}{d_k}(J-1)^{2k} \tag{9.6}$$

Like the Polynomial model, the Ogden model also makes it difficult to determine the material parameters to give an accurate and stable prediction.

9.1.7 Arruda–Boyce Model

The Arruda–Boyce model is based on the eight-chain model that is derived from the deformation behavior of the microstructure of elastomers. The eight-chain model assumes that the chain molecules are positioned along the diagonals of a unit cell in the principal stretch space (see Figure 9.1). The elastic energy density is expressed in terms of μ, limiting network stretch λ_L, and d [1]

$$W = \mu\left[\frac{1}{2}(I_1-3)+\frac{1}{20\lambda_L^2}(I_1^2-9)+\frac{11}{1050\lambda_L^4}(I_1^3-27)\right.$$
$$\left.+\frac{19}{7000\lambda_L^6}(I_1^4-81)+\frac{519}{673750\lambda_L^8}(I_1^5-243)\right] \quad (9.7)$$
$$+\frac{1}{d}\left(\frac{J^2-1}{2}-\ln J\right)$$

The Arruda–Boyce model is more accurate than the Neo-Hookean model and Mooney–Rivlin model, and is nearly as accurate as the Yeoh model. However, it under-predicts the biaxial response because it has no I_2.

Table 9.1 summarizes the predictive capabilities of the listed seven common hyperelastic models, in which the accuracy of the different models is

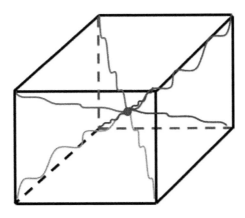

FIGURE 9.1
Eight-chain molecules are positioned in the unit cell.

TABLE 9.1
Comparison between the predictive capabilities of various hyperelasticity models [2]

	Neo-Hookean	Mooney–Rivlin	Yeoh	Arruda–Boyce	Ogden (2-term)	Ogden (3-term)	Gent
R^2-Prediction	0.794	0.843	0.980	0.973	0.977	0.998	0.972

defined by the coefficient of determination (R^2) [2]. The three-term Ogden performs the best; next is the Yeoh. The Arruda–Boyce model, Gent model, and two-term Ogden model are very close behind the first two models. The Neo-Hookean model is the least accurate one because it has the simplest form. The Mooney–Rivlin model is slightly better than the Neo-Hookean model.

9.2 Stability Discussion

Some hyperelastic models are not always stable. For example, the Gent Model under the uniaxial tension in the large strain gets the stress reduced to even negative (see Figure 9.2). Therefore, Drucker's stability should be checked to examine if the model is stable [2]:

$$\Delta(J\sigma) : \Delta(E_{\ln}) \geq 0 \qquad (9.8)$$

where $J = \det[F]$, σ is the Cauchy stress, and $E_{\ln}$ is the logarithmic strain.

Moreover, Drucker's stability should be checked for all loading conditions because some hyperelastic materials may be Drucker's stable in tension but unstable in shear loading. Therefore, in practice, the examination of Drucker's stability of a hyperelastic model is to check the stability of the model in a set of common loading conditions like uniaxial, biaxial, and simple shear. If a material model passes these tests, it may be regarded as relatively safe to use.

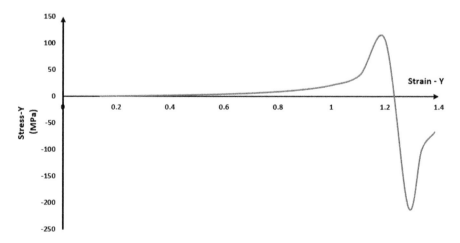

FIGURE 9.2
Stress–strain curve of Gent model under tension.

9.3 Curve-fitting of Material Parameters from Experimental Data

A curve-fitting tool of hyperelasticity is available in ANSYS190. The commands are the same as those for the Chaboche model in Section 3.3, including input of experimental data, definition of material model, initialization of coefficients, and solution. Here is one example of a curve-fitting of the two-term Ogden model:

```
/PREP7
! define uniaxial data
TBFT, EADD, 1, UNIA, CFHY-OG-U.EXP
TBFT, EADD, 1, BIAX, CFHY-OG-B.EXP
! define material: two-term Ogden model
TBFT, FADD, 1, HYPER, Ogden,2
! set the initial value of coefficients
tbft,set,1,hyper,ogde,2,1,1
tbft,set,1,hyper,ogde,2,2,1
tbft,set,1,hyper,ogde,2,3,1
tbft,set,1,hyper,ogde,2,4,1
! define solution parameters if any
TBFT, SOLVE, 1, HYPER, Ogden,2, 1, 500
/OUT
! print the results
TBFT, LIST, 1
tbft,fset,1,hyper,Ogden,2
tblis,all,all
```

The experimental data of rubber for curve-fitting include uniaxial and biaxial tensions [2, 3]. For comparison, the curve-fitting of material parameters was conducted for the Arruda–Boyce model, the three-parameter Mooney–Rivlin model, and the two-term Ogden model. The curve-fitting results are plotted in Figure 9.3, and the obtained material parameters are listed as follows:

Arruda–Boyce model: $\mu = 0.3148$, $\lambda_L = 5.032$

Mooney–Rivlin model: $C_{10} = 1.89e{-}1, C_{01} = -1.933e{-}3, C_{11} = 1.183e{-}4$

Ogden model: $\mu_1 = 3.565e{-}2$, $\alpha_1 = 3.398$, $\mu_2 = 1.272e3$, $\alpha_2 = 4.948e{-}4$

Figure 9.3 indicates that both the Arruda–Boyce model and Ogden model fit the experimental data of the uniaxial tension, while the fitting results of the Mooney–Rivlin model in the uniaxial tension at the large strains are far from the experimental data, which are consistent with the predictive capabilities

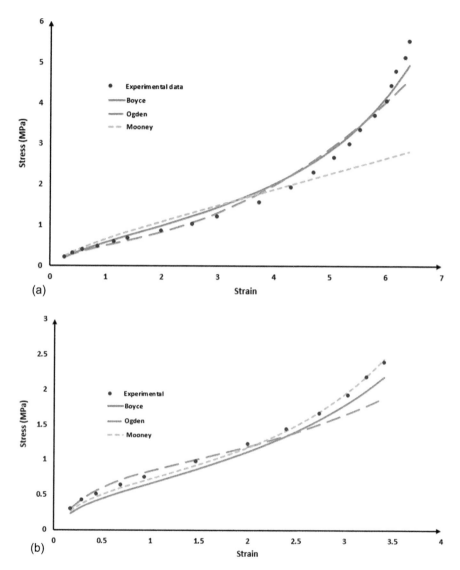

FIGURE 9.3
Curve-fitting results of three hyperelastic models. (a) Uniaxial tension, (b) Biaxial tension.

of various polymers in Table 9.1. For biaxial tension, the Mooney–Rivlin fits well because the strain energy function of the Mooney–Rivlin model has I_2 term, while both the Arruda–Boyce and Ogden models are a little different from the experimental data.

Next, the obtained material parameters were applied to study the rubber rod under compression.

Hyperelasticity

9.4 Simulation of a Rubber Rod under Compression

The study was to simulate the hyperelastic material with material parameters from the experimental data.

9.4.1 Finite Element Model

The rubber rod with a radius of one meter was compressed on the top and deformed within a room enclosed by rigid walls (see Figure 9.4). The rubber rod was meshed with Plane182 (plane strain).

9.4.2 Material Parameters

The curve-fitting results of Section 9.3, including the Arruda–Boyce model, the three-parameter Mooney–Rivlin model, and the two-term Ogden model, were applied to conduct the simulation. Their material models were defined as

```
TB,HYPER,1,,3,MOONEY    !Activate 3 term MOONEY data table
TBDATA,1,1.89e5,-1.933e3,1.183e2

TB,HYPER,1,,1,Boyce     ! Arruda-Boyce
TBDATA,1,3.148e5,5.032

TB, HYPER,1,,2, Ogden   ! two-term Ogden model
TBDATA,1,3.565e4,3.398,1.272e9,4.948e-4
```

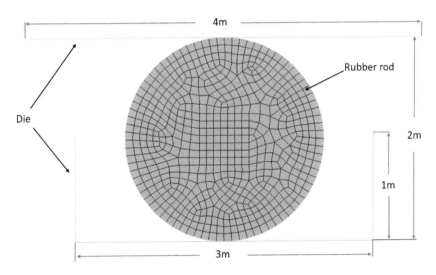

FIGURE 9.4
Finite element model of a rubber rod under compression.

9.4.3 Loadings and Boundary Conditions

The rubber rod was compressed down 0.95 m in the vertical direction. The walls were assumed rigid and constrained in the normal directions (see Figure 9.5).

9.4.4 Results

The von Mises stresses and strains of the rubber rod modeled by the Arruda–Boyce model, the two-term Ogden model, and the Mooney–Rivlin model are depicted in Figures 9.6 through 9.8, respectively. The results of these three models are very similar: the maximum stress and strain occur at the center, and the stresses and strains decrease along the radius from the center to the edge. The maximum stress and strain of these three material models are also very close, around 1.37 MPa for stress and 0.9 for strain.

Figure 9.9 illustrates the strain energy of a rod modeled by the Arruda–Boyce at the end of the loading with a maximum value 0.48 MPa in the center.

9.4.5 Discussion

A 2D rubber rod under compression was modeled in ANSYS190 with material parameters obtained by a curve-fitting of uniaxial and biaxial experimental data. The stresses and strains, as well as the strain energy of the rubber rod, are very consistent.

It is interesting to note that all three material models obtain similar results. Although the curve-fitting results in Section 9.3 show that the Mooney–Rivlin model is far from the uniaxial experimental data in the case of large

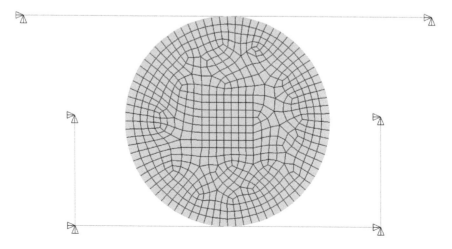

FIGURE 9.5
Boundary conditions.

Hyperelasticity

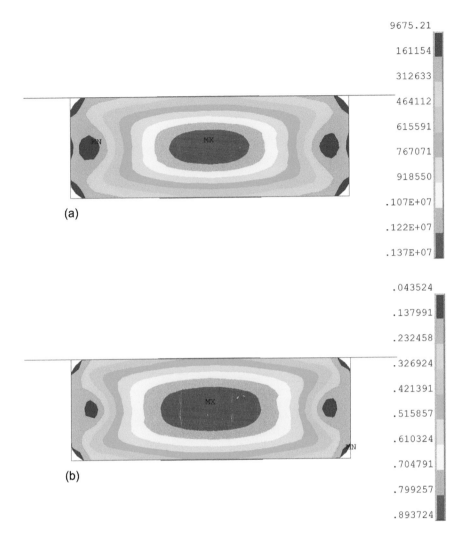

FIGURE 9.6
Results of Arruda–Boyce model. (a) vM stresses (Pa), (b) vM strains.

strain, this model best matches the biaxial data. In the model, the rubber rod is under biaxial compression. That is why all three material models have very similar results.

9.4.6 Summary

The finite element model of a rubber rod was built to study its compression with the material parameters from the curve-fitting. Because the rubber rod is under biaxial compression, all hyperelastic models have very similar results of stresses and strains.

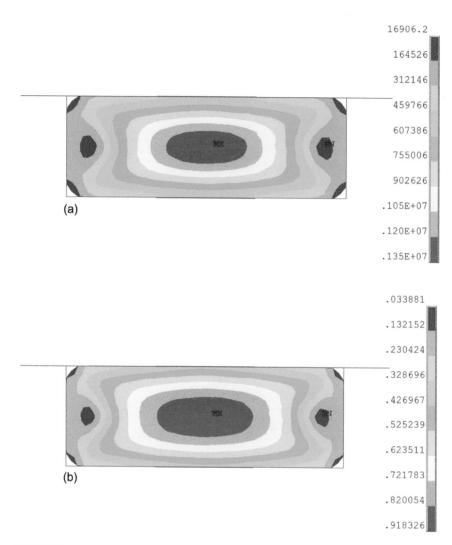

FIGURE 9.7
Results of two-term Ogden model. (a) vM stresses (Pa), (b) vM strains.

9.5 Simulation of Breast Implant in ANSYS

Some women are unhappy with the appearance of their breasts after cancer therapy surgery. Breast implant surgery was developed to address their concerns. In 1998, 122,285 American women selected breast implant surgery [4]. However, clinical studies show that breast implant surgery has safety issues. In addition, the shape of the breast after surgery is another significant

Hyperelasticity

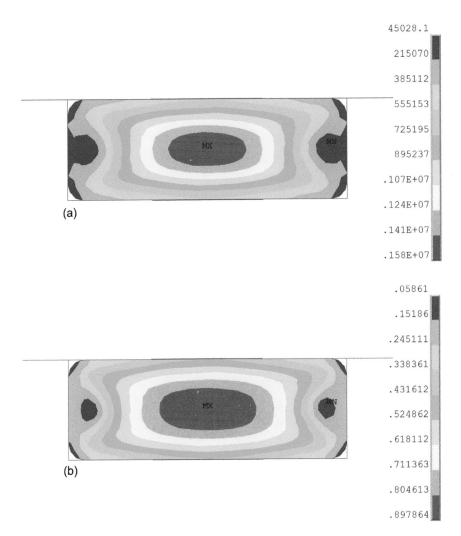

FIGURE 9.8
Results of three-parameter Mooney–Rivlin model. (a) vM stresses (Pa), (b) vM strains.

concern. Therefore, breast implant surgery has been studied intensively, including using the finite element method [5, 6]. Here, we implemented the deformation of a breast under gravity in ANSYS, which can be used for further study of the breast implant.

9.5.1 Finite Element Model

The model is composed of three parts: skin, fat, and glands (see Figure 9.10). The skin was modeled by SHELL181 with thickness 0.7 mm, and the fat and glands were modeled by SOLID187. The skin attached to the fat and glands

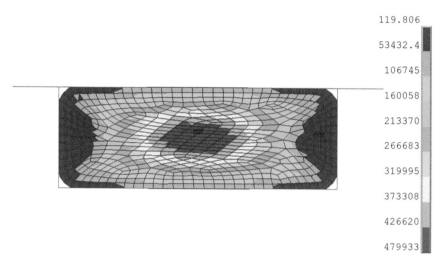

FIGURE 9.9
Strain energy of the rubber rod modeled by Arruda–Boyce material.

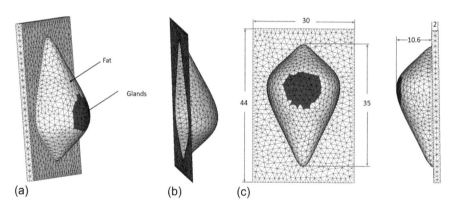

FIGURE 9.10
Finite element model of a breast. (a) Breast model, (b) Skin, (c) Geometry (all dimensions in mm).

share the same nodes. The model can be downloaded from www.feabea.net/models/breast_implant.dat.

9.5.2 Material Models

Although the skin is neither homogeneous nor isotropic, it was simplified as homogeneous and isotropic. In the model, the one-parameter Ogden material model was used to simulate the skin, fat, and glands. They were defined in ANSYS by TB, HYPER and TBDATA commands [7]:

Hyperelasticity

```
TB,HYPER,1,,1,OGDEN      !Activate 1-parameter Ogden for skin
TBDATA,1,803e-6          !Define μ1=803Pa
TBDATA,2,6.8206          !Define α1=6.8206
TB,HYPER,2,,1,OGDEN      !Activate 1-parameter Ogden for Fat
TBDATA,1,1856.7e-6       !Define μ1=1856.7Pa
TBDATA,2,18.1468         !Define α1=18.1468
TB,HYPER,3,,1,OGDEN      !Activate 1-parameter Ogden for
                          glands
TBDATA,1,9065.7e-6       !Define μ1=9065.7Pa
TBDATA,2,25              !Define α1=25
```

The goal of this study is to find the shape of the breast after breast reconstruction; the densities of the fat and glands were given using the TB, DENS command. Here, glands are assumed to have the same density as water.

```
TB,DENS,2                !Density of fat
TBDATA,1,0.75e-6         !Density 0.75g/mm^3
TB,DENS,3                !Density of glands
TBDATA,1,1e-6            !Density 1g/mm^3
```

9.5.3 Loading and Solution Setting

To simulate gravity, 9.8 m/s acceleration was applied on the vertical direction using command ACEL. All material models were defined as hyperelastic materials. Thus, the command for large deflection effect should be turned on. For the boundary conditions, all nodes at the left surface ($y = 2$ mm) were fixed.

9.5.4 Results

Deformation of the whole model shows that under gravity, the breast shape drops 5.25 mm compared to the initial shape just after surgery (see Figure 9.11). Stress concentration occurs at the connection between the body and the breast (see Figure 9.12). The stresses of the skin are much lower than those of the fat and glands, because the skin has the lowest material properties among the three parts.

9.5.5 Discussion

Breast shape under gravity was studied in ANSYS with the Ogden material model for skin and fat. However, the constitutive law of skin is very complicated [8]. A user subroutine of skin is one alternative to model the skin in ANSYS.

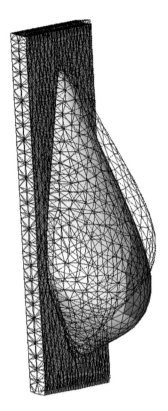

FIGURE 9.11
Deformation of the breast under gravity (DMX=5.25mm).

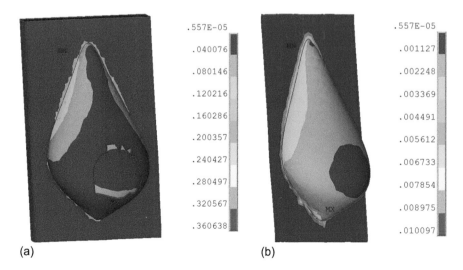

FIGURE 9.12
vM stresses of the breast under gravity (MPa). (a) Breast, (b) Skin.

9.5.6 Summary

Breast shape under gravity was built in ANSYS, and the breast deformation under gravity was plotted after solution. Stress concentration was observed at the connection between the body and the breast.

References

1. ANSYS190 Help Documentation in the help page of product ANSYS190.
2. Bergstrom, J., *Mechanics of Solid Polymers*, Elsevier, 2015.
3. Treloar, L.R., *The Physics of Rubber Elasticity*, Oxford University Press, 1975.
4. Drell, A. (1998, Sept. 6). Breast implants popular again. *Chicago Sun-Times* [On-line], 1–3. Available: http://www.suntimes.com/
5. Wilson, K.A., *Finite Element Analysis of Breast Implants*, Virginia Polytechnic Institute and State University, 1999.
6. Kovar, M., Flynn, C., Sobotka, J., Capek, L., Validation of breast implant finite element model, *Computer Methods in Biomechanics and Biomedical Engineering*, Vol. 20, 2017, pp. 109–110.
7. Samani, A., Bishop, J., Yaffe, M.J., Plewes, D.B., Biomechanical 3-D finite element modeling of the human breast using MRI data, *IEEE Transactions on Medical Imaging*, Vol. 20, 2001, pp. 271–279.
8. Fung, Y.C., *Biomechanics: Mechanical Properties of Living Tissues*, 2nd Edition, Springer, 1993.

10
Viscoelasticity of Polymers

Chapter 10 examines the viscoelasticity of polymers, including various viscoelastic models and their application for study of the viscoelasticity of liver soft tissues.

10.1 Viscoelasticity of Polymers

Polymers exhibit viscoelasticity. For example, when a car runs on the highway, its tires get heated, which is due to the energy dissipation by the viscoelasticity of the tires converted to heat. From the molecular aspect, viscoelasticity is regarded as a molecular rearrangement [1]. When a polymer is loaded with forces, some of the long polymer chains change positions. This rearrangement is named creep. When these long polymer chains move to accompany the loaded forces, the polymers remain a solid material. Thus, a back stress occurs in the material. When the back stress reaches the same level of the applied forces, creep of the material stops. If the applied loadings are taken away, the remaining back stresses cause the polymer to return to its original shape. Some viscoelastic polymers include amorphous polymers, semi-crystalline polymers, and biopolymers.

10.2 Linear Viscoelastic Models

A viscoelastic material has an elastic part and a viscous part. The elastic part can be modeled as springs, and its stress–strain relationship is expressed as

$$\sigma = E\varepsilon \qquad (10.1)$$

The viscous part can be regarded as dashpots, and its stress is determined by its time derivative of strain

$$\sigma = \eta \frac{d\varepsilon}{dt} \qquad (10.2)$$

where η is the viscosity of the material.

Some models, such as the Maxwell model, the Kelvin–Voigt model, and the Burgers model, have been developed to predict the response of viscoelastic material under different loading conditions [2]. These models contain different combinations of springs and dashpots, which are briefly introduced as follows.

10.2.1 Maxwell Model

In the Maxwell model, a viscous damper and an elastic spring are connected in a series (see Figure 10.1). The model is described by the following equation:

$$\sigma + \frac{\eta}{E}\dot{\sigma} = \eta\dot{\varepsilon} \tag{10.3}$$

The above equation indicates that under a constant strain, the stresses decay gradually. Thus, the Maxwell model accurately predicts the stress relaxation of the polymers. However, the equation (10.3) shows that under a constant stress, the strain increases linearly with time as long as the force is loaded, which goes against the mechanical behavior of most polymers. Therefore, the Maxwell model cannot be used for modeling creep.

10.2.2 Kelvin–Voigt Model

The Kelvin–Voigt model can be represented by a viscous damper and an elastic spring connected in parallel (see Figure 10.2). The model is expressed by

$$\sigma = E\varepsilon + \eta\dot{\varepsilon} \tag{10.4}$$

The Kelvin–Voigt model shows that under a constant stress, the deformation rate of the material reduces gradually until the material reaches a steady state. After the stress is removed, the material returns gradually to its undeformed state. Thus, unlike the Maxwell model, the Kelvin–Voigt model can predict the creep well. Yet, it models the stress relaxation much less accurately than the Maxwell model.

FIGURE 10.1
Maxwell model.

Viscoelasticity of Polymers

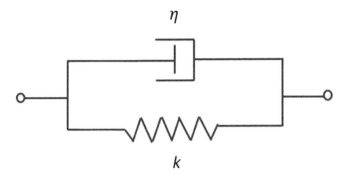

FIGURE 10.2
Kelvin–Voigt model.

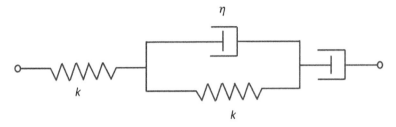

FIGURE 10.3
Burgers model.

10.2.3 Burgers Model

In order to overcome the limitations of the Maxwell model and Kelvin–Voigt model, the Burgers model was created to combine the Maxwell and Kelvin–Voigt models in a series (see Figure 10.3). The constituent relation is given as

$$\sigma + \left(\frac{\eta_1}{E_1} + \frac{\eta_2}{E_1} + \frac{\eta_2}{E_2}\right)\dot{\sigma} + \frac{\eta_1}{E_1}\frac{\eta_2}{E_2}\ddot{\sigma} = \eta_2\dot{\varepsilon} + \frac{\eta_1\eta_2}{E_1}\ddot{\varepsilon} \qquad (10.5)$$

The above second-order differential equation of the Burgers model can be used to describe the viscoelastic behaviors of the polymers under different loadings.

10.2.4 Generalized Maxwell Model

The relaxation of the polymers has a varying time distribution because shorter molecular segments contribute less than longer ones. Therefore, the Generalized Maxwell model has many spring-dashpot Maxwell elements that consider this factor (see Figure 10.4).

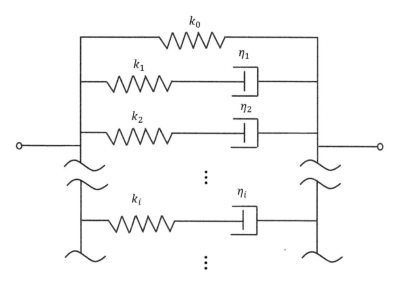

FIGURE 10.4
Generalized Maxwell model.

A Generalized Maxwell model in three dimensions is described by the Prony series

$$\sigma = \int_0^t 2G(t-\tau)\frac{de}{d\tau}d\tau + I\int_0^t K(t-\tau)\frac{d\Delta}{d\tau}d\tau \qquad (10.6)$$

where $G(t)$ – the Prony series shear-relaxation moduli, expressed as

$$G(t) = G_0\left[a_\infty^G + \sum_{i=1}^{n_G} a_i^G \exp\left(-\frac{t}{\tau_i^G}\right)\right] \qquad (10.7)$$

$K(t)$ – the Prony series bulk-relaxation moduli, specified by

$$K(t) = K_0\left[a_\infty^K + \sum_{i=1}^{n_K} a_i^K \exp\left(-\frac{t}{\tau_i^K}\right)\right] \qquad (10.8)$$

G_0, K_0 – relaxation moduli at $t = 0$;
n_G, n_K – number of Prony terms;
a_i^G, a_i^K – relative moduli;
τ_i^G, τ_i^K – relaxation time;
σ – Cauchy stress;
e – deviatoric strain;
Δ – volumetric strain;
τ – time; and
I – identity tensor.

Viscoelasticity of Polymers

In ANSYS, Prony series constants are specified by [3]

```
MP, EX, 1, E               ! define Young's modulus
MP, NUXY, 1, ν             ! define Poisson's ratio
TB, PRONY, , , NG, SHEAR   ! the shear Prony data table
TBDATA,1, α₁ᴳ, τ₁ᴳ, ... αₙᴳ, τₙᴳ,
TB, PRONY, , , NK, BULK    ! the bulk Prony data table
TBDATA,1, α₁ᵏ, τ₁ᵏ, ... αₙᵏ, τₙᵏ,
```

Next, the Prony series was applied to study the viscoelasticity of liver soft tissues.

10.3 Viscoplasticity Models

Viscoplasticity models can predict the response of nonlinear, time-dependent polymers, but they require more experimental data to define material parameters. The Bergstrom–Boyce model is one typical viscoplasticity model, which is introduced in this section.

The Bergstrom–Boyce model can be represented using two parallel networks, A and B (see Figure 10.5) [4]. A is a nonlinear hyperelastic component, and B is a nonlinear hyperelastic component in a series with a nonlinear viscoelastic flow part. The total stress is expressed by

$$\sigma = \sigma_A + \sigma_B \tag{10.9}$$

$$\sigma_A = \frac{\mu}{J\bar{\lambda}^*} \frac{\Gamma^{-1}\left(\bar{\lambda}^*/\lambda_{lock}\right)}{\Gamma^{-1}\left(1/\lambda_{lock}\right)} \mathrm{dev}\left[b^*\right] + \kappa(J-1)I \tag{10.10}$$

$$\sigma_B = \frac{s\mu}{J_B^e \bar{\lambda}_B^{e*}} \frac{\Gamma^{-1}\left(\bar{\lambda}^{e*}/\lambda_{lock}\right)}{\Gamma^{-1}\left(1/\lambda_{lock}\right)} \mathrm{dev}\left[b_B^{e*}\right] + \kappa(J_B-1)I \tag{10.11}$$

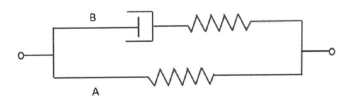

FIGURE 10.5
Bergstrom–Boyce model.

where μ, λ^{lock}, κ are material parameters, Γ^{-1} is the inverse Langevin function, b^* is the distortional part of left Cauchy–Green tensor, s is a dimensionless material parameter that specifies the stiffness of network B over that of network A, and $\overline{\lambda^{e*}}$ is the chain stretch of network B.

The rate-equation for viscous flow is specified by [4]

$$\dot{\gamma}_B^v = \dot{\gamma}_0 (\overline{\lambda_B^v} - 1 + \xi)^C \left[R\left(\frac{\tau}{f_v \tau_{base}} - \hat{\tau}_{cut} \right) \right]^m \quad (10.12)$$

where ξ, C, τ_{base}, $\hat{\tau}_{cut}$, m are material parameters, $\dot{\gamma}_0$ is a constant to ensure dimensional consistency, $R(x)$ is the ramp function, and $\overline{\lambda_B^v}$ is the viscoelastic chain stretch.

The Bergstrom–Boyce model is widely used to predict the nonlinear time-dependent large-strain behavior of elastomer-like materials.

10.4 Simulation of Viscoelasticity of Liver Soft Tissues

In vehicle crashes, the liver is the most often injured abdominal organ [5, 6]. In the previous study, the liver was always modeled with hyperelastic materials [7–12]. In this study, the viscoelasticity of the liver under compression was simulated in ANSYS190.

10.4.1 Finite Element Model

The liver lay on a plate and was compressed by a cylinder on the top. The whole model was meshed with Plane182 in plane strain condition (see Figure 10.6).

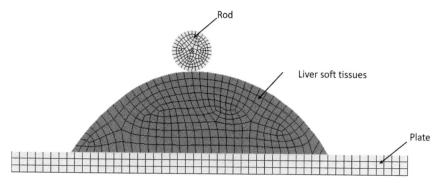

FIGURE 10.6
Finite element model of testing of liver soft tissues.

10.4.2 Material Properties

The liver was modeled with the one-term Ogden model, and its viscoelasticity was simulated with the Prony Series. The material parameters, which are listed in Table 10.1, were defined by ANSYS APDL commands as

```
TB,HYPE,1,1,1,OGDEN
TBDATA,1,3.7e-2,5.59,
TB,PRONY,1,1,3,SHEAR
tbdata,1,6.9701e-3,10,5.83e-2,100,3.53e-2,1000
```

10.4.3 Contact Definition

The contact between the plate and the liver and the contact between the liver and the cylinder were assumed as a standard contact. The cylinder was regarded as rigid and controlled by a pilot node.

10.4.4 Loadings and Boundary Conditions

The loading is divided into two steps. In the first step, the cylinder moved down 0.5mm in 0.1 second. In the second step, it held 1000s. The bottom of the plate was fixed in the vertical direction (see Figure 10.7).

TABLE 10.1
Material parameters of the liver [13]

	a_n^G (MPa)	τ_n^G
1	6.9701e−3	10
2	5.83e−2	100
3	3.53e−2	1000

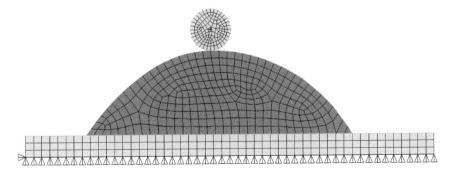

FIGURE 10.7
Boundary conditions.

10.4.5 Results

The von Mises stresses of the model at the end of the first and second steps are presented in Figures 10.8 and 10.9, respectively. The stresses are localized around the loading area, and the maximum stress decrease was around 10% between the first step and second step. Figure 10.10 illustrates the time history of the reaction force, which decreases with time and then approaches a constant after 600s.

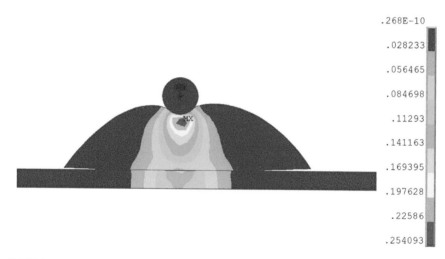

FIGURE 10.8
vM stresses of soft tissues at the end of the first step (t=0.1s) (MPa).

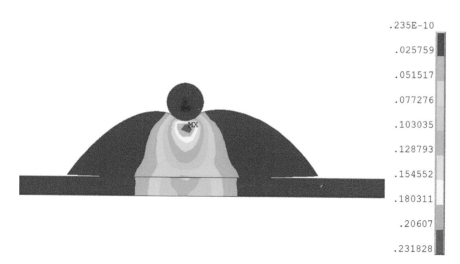

FIGURE 10.9
vM stresses of soft tissues at the end of the second step (t=1000s) (MPa).

Viscoelasticity of Polymers

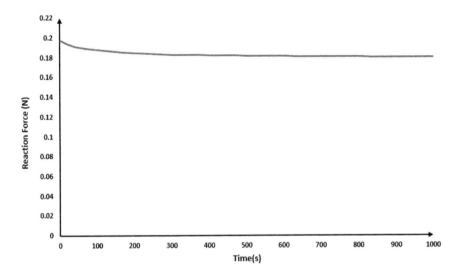

FIGURE 10.10
Reaction force with time.

10.4.6 Discussion

A 2D finite element model of the liver was built to study its viscoelastic response under the compression of a cylinder. The reaction force decreases with time and approaches a constant after a certain time, which is because the elastic Ogden model controls its stress response in the first loading step, and the Prony series governs its viscoelastic behavior. Without a definition of the Prony series in the material model, the stress state and reaction force should not change in the second step.

The drawback of this study is that the liver was simplified as plane strain state, which significantly differs from the true liver. Future studies should use a 3D model of the liver with the geometry from a CT scan.

10.4.7 Summary

The viscoelasticity of the liver under compression of a cylinder was studied in ANSYS190. The stress and reaction force decrease with time, which reflects the viscoelasticity of the liver.

References

1. McCrum, N.G., Buckley, C.P., Bucknell, C.B., *Principles of Polymer Engineering*, Oxford University Press, 2003.
2. Roylance, D., *Engineering Viscoelasticity*, MIT, 2001.

3. ANSYS190 Help Documentation in the help page of product ANSYS190.
4. Bergstrom, J., *Mechanics of Solid Polymers*, Elsevier, 2015.
5. Elhagediab, A.M., Rouhana, S., Patterns of abdominal injury in frontal automotive crashes, 16th International ESV Conference Proceedings NHTSA Washington DC, 1998, pp. 770–780.
6. Rouhana, S.W., Foster, M.E., Lateral impact-an analysis of the statistics in the NCSS, Proc. 29th Stapp Car Crash Conference, 1985, pp. 79–98.
7. Gao, Z., Lister, K., Desai, J.P., Constitutive modeling of liver tissue: experiment and theory, *Annals of Biomedical Engineering*, Vol. 38, 2010, pp. 505–516.
8. Roan, E., Vemaganti, K., The nonlinear material properties of liver tissue determined from no-slip uniaxial compression experiments, *Journal of Biomechanical Engineering*, Vol. 129, 2007, pp. 450–456.
9. Chui, C., Kobayashi, E., Chen, X., Hisada, T., Sakuma, I., Combined compression and elongation experiments and non-linear modeling of liver tissue for surgical simulation, *Medical & Biological Engineering & Computing*, Vol. 42, 2004, pp. 787–798.
10. Sakuma, I., Nishimura, Y., Chui, C., Kobayashi, E., Inada, H., Chen, X., Hisada, T., In vitro measurement of mechanical properties of liver tissue under compression and elongation using a new test piece holding method with surgical glue, *Surgery Simulation and Soft Tissue Modeling*, Ayache, N. and Delingette, H. (eds.), pp. 284–292, Springer, 2003.
11. Fu, Y.B., Chui, C.K., Teo, C.L., Liver tissue characterization from uniaxial stress-strain data using probabilistic and inverse finite element methods, *Journal of the Mechanical Behavior of Biomedical Materials*, Vol. 20, 2013, pp. 105–112.
12. Umale, S., Deck, C., Bourdet, N., Dhumane, P., Soler, L., Marescaux, J., Willinger, R., Experimental mechanical characterization of abdominal organs: liver, kidney & spleen, *Journal of the Mechanical Behavior of Biomedical Materials*, Vol. 17, 2013, pp. 22–33.
13. Sato, F., Yamamoto, Y., Ito, D., Antona-Makoshi, J., Ejima, S., Kamiji, K., Yasuki, T., Hyper-viscoelastic response of perfused liver under dynamic compression and estimation of tissue strain thresholds with a liver finite element model, IRCOBI Conference, 2013, pp. 736–750.

11

Mullins Effect

Chapter 11 first introduces the Mullins effect and then focuses on the Ogden–Roxburgh Mullins effect model. The final part of Chapter 11 shows the damage evolution of rubber tires on the road in ANSYS.

11.1 Introduction of Mullins Effect

Significant softening of the elastomers occurs at the first few loading cycles (see Figure 11.1). After that, the material properties remain constant [1–4]. This phenomenon is called the Mullins effect. The Mullins effect happens only to the elastomers and elastomer-like materials. From the aspect of the molecular structure, the damage is caused by either a break of molecular chains at the interface of filler particles or a breakdown of filler particle clusters. Experiments find the following features of the Mullins effect:

(1) Softening increases with filler particle concentration;
(2) In the first few cycles, stress to a given strain is higher than in the following cycles;
(3) Damage in the material becomes the highest in the first few cycles and reduces significantly in the following cycles;
(4) Maximum applied strain determines the amount of damage. When the damage is caused by a cyclic loading, the material exhibits more damage accumulation if the higher loading is applied in the following cycles; and
(5) The Mullins effect can be recovered slowly with time, and temperature significantly affects its recovery rate.

A few models have been developed to predict the Mullins effect such as the Ogden–Roxburgh Mullins effect model and Qi–Boyce Mullins effect model [5]. The Ogden–Roxburgh Mullins effect model, which was implemented in ANSYS, is introduced in the following section.

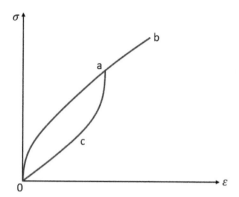

FIGURE 11.1
Response of Mullins effect model. (a) First loading 0→a→b (b) First unloading b→a→c→0 (c) Second loading 0→c→a.

11.2 Ogden–Roxburgh Mullins Effect Model

In the Ogden–Roxburgh Mullins effect model, an internal variable η is introduced in the Helmholtz free energy density Ψ to track the damage evolution in the material, which is expressed as [6]

$$\eta = 1 - \frac{1}{r}\mathrm{erf}\left(\frac{\psi^{max} - \psi}{m + \beta\psi^{max}}\right) \qquad (11.1)$$

where
$r, \beta,$ and m are material constants,
erf (x) – the error function,
ψ – current strain energy density for the material point, and
ψ^{max} – maximum strain energy density for the material point in its loading history.

The Ogden–Roxburgh Mullins effect model can be applied to predict any isotropic hyperelastic material model.
The Mullins effect is defined in ANSYS as

```
TB, CDM, , , , PSE2
TBDATA, 1, r, m, β
```

11.3 Simulation of a Rubber Tire with the Mullins Effect

In this study, a tire compressed on a rigid road was modeled in ANSYS accompanying the Mullins effect. The damage during the rotation of the

Mullins Effect

tire should be helpful for understanding the physical meaning of the Mullins effect.

11.3.1 Finite Element Model

The model was created to represent a tire on a rigid road that was simplified as a rigid plate (see Figure 11.2). The tire was meshed by SOLID185.

11.3.2 Material Properties

The rubber was simulated with a three-term YEOH material model. In addition, the damage of the rubber was modeled using the modified Ogden–Roxburgh Mullins effect. The corresponding APDL commands are given as

```
! material properties (Yeoh)
TB,HYPER,1,,3,YEOH    !Activate 3-term Yeoh data table
TBDATA,1, 1.317e6     !Define C1
TBDATA,2, 0.0341e6    !Define C2
TBDATA,3, 0.0037e6    !Define C3
TBDATA,4,3.0E-9       !Define first incompressibility parameter

TB,CDM,1,,3,PSE2      !modified Ogden-Roxburgh Mullins effect
TBDATA,1,2.34, 0.069e6,0.222
```

11.3.3 Loadings and Boundary Conditions

The loading was completed in two steps. In the first step, the rigid plate moved up 0.2 m to compress the tire (see Figure 11.3a). In the second step, the tire rotated three revolutions under compression (see Figure 11.3b).

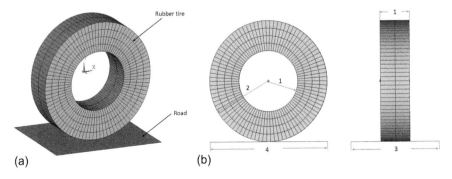

FIGURE 11.2
Finite element model of a rubber tire on the road. (a) Finite element model (b) Geometry (all dimensions in meters).

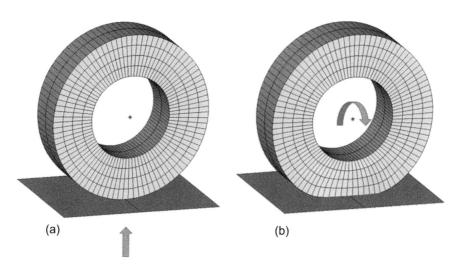

FIGURE 11.3
Loadings. (a) Plate moves up to compress the rubber tire (b) Rubber tire rotates three rounds.

11.3.4 Results

Figure 11.4 shows the deformation and the stresses of the tire under compression at the end of the first step. The stresses of the tire in the second loading steps, which are very close to each other, but lower than that at the end of the first step, are presented in Figure 11.5. The damage in the first revolution is plotted in Figure 11.6, and at the end of the second and third revolution in Figures 11.7 and 11.8, respectively. Figures 11.9 through 11.12 illustrate the reaction forces and reaction moments with time and revolutions. Both show that the reaction force and moment have a sharp drop at the end of the first revolution, which is consistent with the damage change.

11.3.5 Discussion

A 3D finite element model of a tire compressed on a rigid road was developed to simulate the Mullins effect of the tire. The variation of the damage variable with the rotation of the tire is consistent with the reaction force and reaction moment with time and revolution of the tire.

The damage variable in the Mullins effect differs from other damage models such as the composite damage in Chapter 19. The damage variable in the Mullins effect is associated with the virgin material strain-energy potential W_0 and the maximum virgin potential over the time W_m. If $W_0 = W_m$, the damage variable becomes 1. If $W_0 < W_m$, the damage variable is less than 1. This explains why the red color referring to 1 in the damage contour disappears

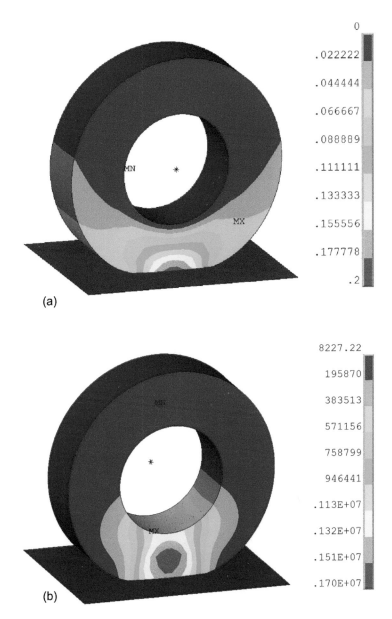

FIGURE 11.4
Tire under compression at the end of the first step. (a) Deformation (b) vM stresses (Pa).

gradually in the first revolution. After the first revolution, both W_0 and W_m are stable; that is why the reaction force and moment have a slight change in the second and third revolutions. In the Mullins effect, the physical meaning of the damage variable looks more like the effect variable than the damage variable.

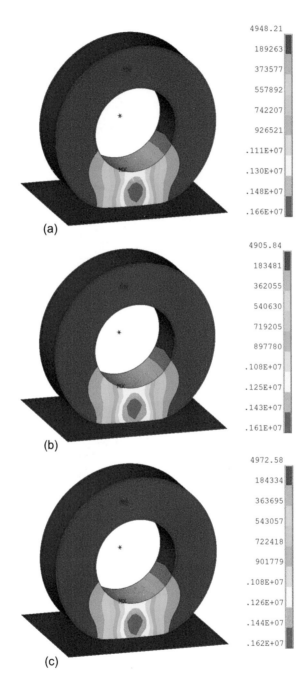

FIGURE 11.5
Stresses of the tire in the second loading steps (Pa). (a) End of first revolution (b) End of second revolution (c) End of third revolution.

Mullins Effect

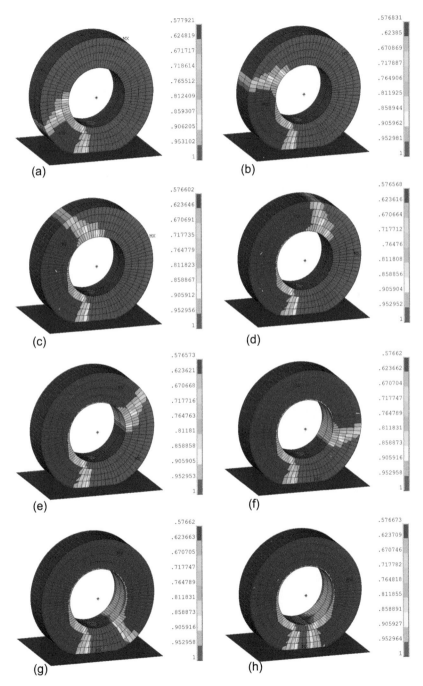

FIGURE 11.6
Damage in the first revolution. (a) Rotate 45° (b) Rotate 90° (c) Rotate 135° (d) Rotate 180° (f) Rotate 225° (f) Rotate 270° (g) Rotate 315° (h) Rotate 360°.

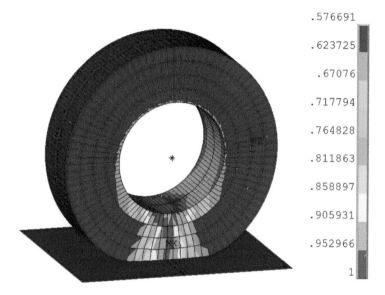

FIGURE 11.7
Damage of the rubber tire at the end of the second revolution.

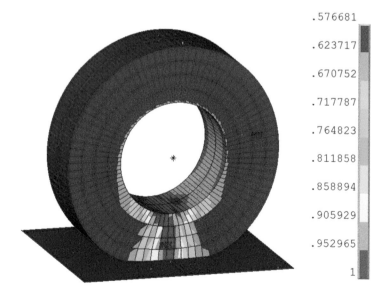

FIGURE 11.8
Damage of the rubber tire at the end of the third revolution.

Mullins Effect

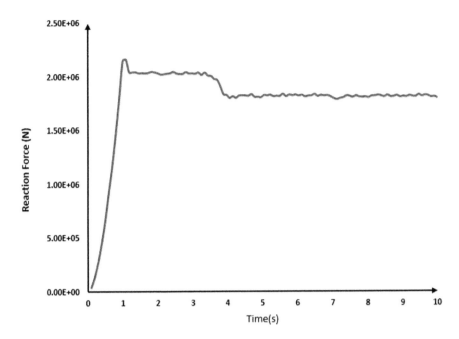

FIGURE 11.9
Reaction force with time.

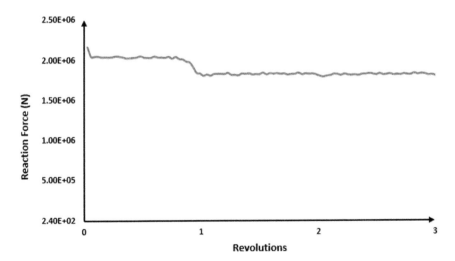

FIGURE 11.10
Reaction force with cycle number.

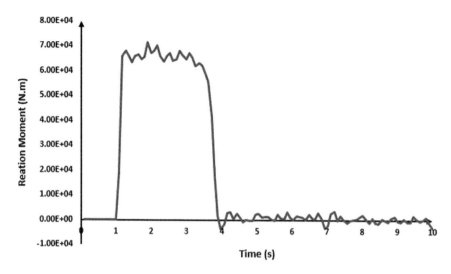

FIGURE 11.11
Reaction moment with time.

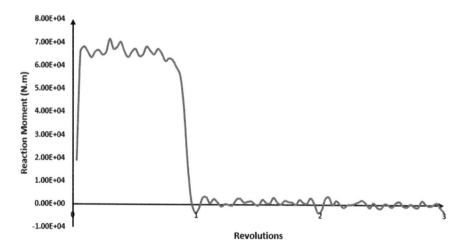

FIGURE 11.12
Reaction moment with cycle number.

11.3.6 Summary

The Mullins effect of a tire compressed on a rigid road was studied in ANSYS190. The reaction force and moment of the tire match the evolution of the damage variable.

References

1. Harwood, J.A.C., Mullins, L., Payne, A.R., Stress softening in natural rubber vulcanizates. Part II Stress softening in pure gum and filler loaded rubbers, *Journal of Applied Polymer Science*, Vol. 9, 1965, pp. 3011–3021.
2. Harwood, J.A.C., Payne, A.R., Stress softening in natural rubber vulanizates. Part III Carbon black-filled vulanizates, *Journal of Applied Polymer Science*, Vol. 10, 1966, pp. 315–324.
3. Harwood, J.A.C., Payne, A.R., Stress softening in natural rubber vulcanizates. Part IV Unfilled vulcanizates, *Journal of Applied Polymer Science*, Vol. 10, 1966, pp. 1203–1211.
4. Diani, J., Fayolle, B., Gilormini, P., A review of Mullins effects, *European Polymer Journal*, Vol. 45, 2009, pp. 601–612.
5. Bergstrom, J., *Mechanics of Solid Polymers*, Elsevier, 2015.
6. Ogden, R.W., Roxburgh, D.G., A pseudo-elastic model for the Mullins effect in filled rubber, *Proceedings of the Royal Society of London, Series A*, Vol. 455, 1999, pp. 2861–2877.

12

Usermat for Hyperelastic Materials

ANSYS provides a powerful tool for customers to develop their own hyperelastic models. This is UserHyper in ANSYS190, which is introduced in Chapter 12 along with one example to reproduce the Gent model.

12.1 Introduction of Subroutine UserHyper

Subroutine UserHyper in ANSYS follows the format as [1]:

```
subroutine UserHyper(
    &           prophy, incomp, nprophy, invar,
    &           potential, pInvDer)
```

The above input and output parameters are explained as follows:

nrophy – number of material parameters,

prophy – array of material parameters,

invar – invariants I_1, I_2, and J,

potential – strain energy W, and

pInvDer – derivative of the strain-energy potential with respect to I_1, I_2, and J.

pInvDer includes nine terms $\partial W/\partial I_1$, $\partial W/\partial I_2$, $\partial^2 W/\partial I_1 \partial I_1$, $\partial^2 W/\partial I_1 \partial I_2$, $\partial^2 W/\partial I_2 \partial I_2$, $\partial^2 W/\partial I_1 \partial J$, $\partial^2 W/\partial I_2 \partial J$, $\partial W/\partial J$, and $\partial^2 W/\partial J \partial J$ consecutively from one to nine.

12.2 Simulation of Gent Hyperelasticity

As an example, subroutine UserHyper was applied to simulate the Gent hyperelastic material model.

12.2.1 Subroutine UserHyper for Gent Material

The strain energy density function of the Gent hyperelasticity has the following form [2]:

$$W(I_1) = -\frac{\mu J_m}{2}\ln\left(1-\frac{I_1-3}{J_m}\right)+\frac{1}{d}\left(\frac{J^2-1}{2}-\ln J\right) \quad (12.1)$$

Calculating the first and second derivatives of the strain-energy potential with respect to the three invariants

$$\frac{\partial W}{\partial I_1} = \frac{\mu J_m}{2(J_m-I_1+3)} \quad (12.2)$$

$$\frac{\partial W}{\partial J} = \frac{1}{d}\left(J-\frac{1}{J}\right) \quad (12.3)$$

$$\frac{\partial^2 W}{\partial I_1 \partial I_1} = \frac{\mu J_m}{2(J_m-I_1+3)^2} \quad (12.4)$$

$$\frac{\partial^2 W}{\partial J \partial J} = \frac{1}{d}\left(1+\frac{1}{J^2}\right) \quad (12.5)$$

The other derivatives are zero. Thus, the above was implemented in the UserHyper subroutine.

```
      Subroutine UserHyper(
     &             prophy, incomp, nprophy, invar,
     &             potential, pInvDer)
#include "impcom.inc"
      DOUBLE PRECISION ZERO, ONE, TWO, THREE, HALF, TOLER
      PARAMETER   (ZERO = 0.d00,
     &             ONE   = 1.0d0,
     &             HALF = 0.5d0,
     &             TWO   = 2.d0,
     &             THREE = 3.d0,
     &             TOLER = 1.0d-12)
c
c --- argument list
c
      INTEGER    nprophy
      DOUBLE PRECISION prophy(*), invar(*),
     &             potential, pInvDer(*)
      LOGICAL    incomp
c
c --- local variables
```

```
c
  c
        DOUBLE PRECISION i1, jj, u, jm, oD1, j1
c
        i1  = invar(1)
        jj  = invar(3)
        u   = prophy(1)
        jm  = prophy(2)
        oD1 = prophy(3)
        potential  = ZERO
        pInvDer(1) = ZERO
        pInvDer(3) = ZERO
        potential  = -u*jm/TWO*log(1-(i1-3)/jm)
        pInvDer(1) = u*jm/TWO/(jm-i1+3)
        pInvDer(3) = pInvDer(1)/(jm-i1+3)

        j1 = ONE / jj
        pInvDer(8) = ZERO
        pInvDer(9) = ZERO
        IF(oD1 .gt. TOLER) THEN
          oD1 = ONE / oD1
          incomp = .FALSE.
          potential = potential + oD1*((jj*jj - ONE)*HALF
     -              - log(jj))
          pInvDer(8) = oD1*(jj - j1)
          pInvDer(9) = oD1*(ONE + j1*j1)
        END IF
c
        RETURN
        END
```

12.2.2 Validation

A uniaxial tension test was conducted with equivalent material properties defined by tb, hyper, „ ,user and tb, hyper, „ ,gent, respectively.

```
! define material by tb, hyper, ,, ,user
TB, HYPER, 1, ,3, USER
TBDATA, 1, 1.5, 10, 1e-3
! define material by tb, hyper, ,, ,Gent
TB, HYPER, 1, ,3, Gent
TBDATA, 1, 1.5, 10, 1e-3
```

The stresses obtained by tb,hyper, „ ,user and tb,hyper, „ ,Gent match very well (see Figure 12.1), which validates the developed UserHyper.

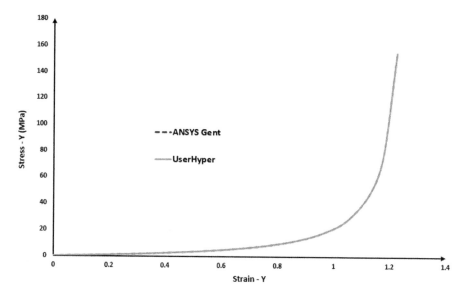

FIGURE 12.1
Validation of UserHyper using ANSYS Gent model.

12.2.3 Summary

The Gent model was reproduced by subroutine UserHyer with less than a 20-line code. Compared with a two-page code using the traditional usermat to simulate the Neo-Hookean model [2], subroutine UserHyper is demonstrated to be simple and useful for the simulation of hyperelastic models.

References

1. ANSYS190 help documentation of the help page of product ANSYS190.
2. Bergstrom, J., *Mechanics of Solid Polymers*, Elsevier, 2015.

Part III

Soil

When we construct a big building such as a skyscraper, we must analyze the foundation to ensure the building is stable after its completion. Similarly, when we do a tunnel excavation, we also need to engage in structural analysis. Works like these can benefit from the finite element analysis, which involves various material models of the soil. This is the topic of Part III.

Chapter 13 presents a brief introduction to soil. Then, Chapters 14 to 17 focus on four major material models of soil and their applications, including the Cam Clay model, Drucker–Prager model, Mohr–Coulomb model, and Jointed Rock model.

Soil consists of a solid phase and water. When the soil is under compression, it expels water. That is the soil consolidation discussed in Chapter 18.

13

Soil Introduction

Chapter 13 primarily highlights soil structure and the parameters that characterize the soil.

13.1 Soil Structure

Sand, silt, and clay are three main mineral components of soil. Their relative proportions control a soil's texture (see Figure 13.1), and the soil's texture influences the properties of the soil such as porosity, permeability, and water-holding capacity. Sand and silt come from the physical and chemical weathering of the parent rock [1], while clay is produced by the precipitation of the dissolved parent rock. Sand is the least active; clay is the most active, followed by silt. Sand helps the soil resist compaction and increase soil porosity. Silt has a higher specific surface area than sand, which makes silt more chemically and physically active than sand. Clay, due to its very specific surface area and large number of negative charges, enables the soil to retain water and nutrients. From the particle size, sand consists mainly of quart particles with a diameter that ranges from 2.0 to 0.04 mm, silt ranges from 0.05 to 0.002 mm, and clay is as small as 0.002 mm. Soil with a size over 2.0 mm is classified as rock and gravel.

13.2 Soil Parameters

Soil is characterized by some parameters, including the following

(1) Density. Soil particle density ranges from 2.60 to 2.75 g/cm^3. The soil bulk density is defined as (dry mass of soil/volume of soil), which is always smaller than soil particle density. Thus, the soil bulk density can be used for characterizing soil compaction.

(2) Porosity. Soils are composed of solid particles. Voids exist between the particles, in which air, water, or both stay. Porosity is calculated by 1 − (bulk density/particle density).

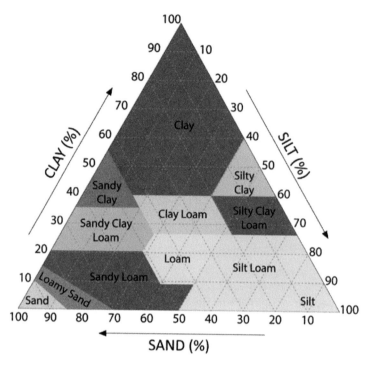

FIGURE 13.1
Soil type (Kanda Euatham © 123rf.com).

(3) Consistency. Consistency refers to the cohesion and adhesion ability of soil, and its ability to sustain deformation and rupture. It is very useful for predicting cultivation problems and the engineering of foundations [2]. In civil engineering, soil consistency is used prior to construction for evaluating soil that supports buildings and roads.

References

1. Jenny, H., *Factors of Soil Formation: A System of Quantitative Pedology*, McGraw-Hill, 1941.
2. Merritt, F.S., Rickett, J.T., *Building Design and Construction Handbook*, 6th Edition, McGraw-Hill Professional, 2000.

14

Cam Clay Model

Chapter 14 focuses on the Cam Clay model. Sections 14.1 and 14.2 introduce the Modified Cam Clay model, including its advantages and how to define the Cam Clay Model in ANSYS. In addition, the subsidence of a tower is simulated in Section 14.3.

14.1 Introduction of Modified Cam Clay Model

The Cam Clay model and modified Cam Clay model were presented first by the researchers at Cambridge University to describe the behaviors of soft soils [1, 2]. The modified Cam Clay model was developed from the Cam Clay model; it has been widely used in geotechnical engineering since predication of soil deformation is crucial in many geotechnical engineering problems. Thus, the following introduction is limited to the modified Cam Clay model.

In the modified Cam Clay model, the voids between the solid particles are assumed to be full of only water. When the soil is loaded, the water is expelled from the voids, and the volume of the soil changes. The modified Cam Clay yield function is an elliptic function [3]

$$\frac{q^2}{p'^2} + M^2\left(1 - \frac{p'_c}{p'}\right) = 0 \tag{14.1}$$

In the p'-q plane, the modified Cam Clay yield function is an ellipse (see Figure 14.1).

The modified Cam Clay model can predict unlimited soil deformations (including softening and hardening) with a constant stress or volume when a critical state is reached. Figure 14.2 illustrates the softening behavior of the modified Cam Clay model. Before the effective stress path (dash line) touches the initial yield surface, the heavily over-consolidated soil is elastic. After the effective stress path touches the initial yield surface, the soil starts to contract (softening); it undergoes plasticity until the effective stress path reaches F point where the failure occurs. The hardening behavior of the modified Cam Clay model is depicted in Figure 14.3. Within the initial yield surface, the normally consolidated and lightly over-consolidated soil is elastic. After the effective stress path reaches the initial yield surface, the soil

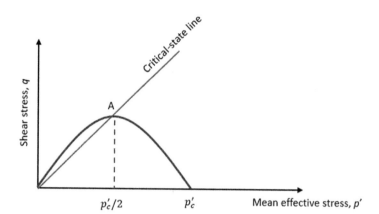

FIGURE 14.1
Yield function of the modified Cam Clay model.

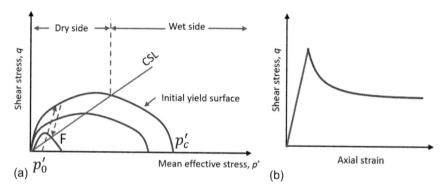

FIGURE 14.2
Softening behavior of the modified Cam Clay model [3]. (a) Development of a yield surface (dash line). (b) Strain–stress curve.

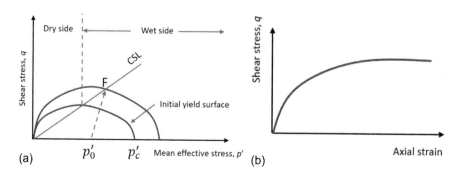

FIGURE 14.3
Hardening behavior of the modified Cam Clay model [3]. (a) Development of a yield surface (dash line). (b) Strain–stress curve.

Cam Clay Model

has plastic strains. The yield surface becomes larger (hardening) until the effective stress path reaches point F where the soil fails.

Another advantage of the modified Cam Clay model is that the yield surface (see Figure 14.1) is very smooth, which makes the solution converged easily. As a result, the modified Cam Clay has wide applications for geotechnical engineering.

14.2 Cam Clay Model in ANSYS

The Cam Clay model in ANSYS consists of the elastic component and plastic component, which are individually introduced.

14.2.1 Elastic Component

The stiffness matrix C is composed of the bulk modulus and shear modulus [4],

$$C = k_b I \otimes I + 2\mu I_{dev} \tag{14.2}$$

where
k_b – bulk modulus;
μ – shear modulus;
I – second-order identity tensor; and
I_{dev} – the fourth-order deviatoric projection tensor.

During the compression of the soil, the bulk modulus is a function of the elastic volumetric strain.

$$k_b = \left(p + p_t^{el}\right)\left(\frac{1+e_0}{\kappa} e^{\varepsilon_{vol}^{el}}\right) \tag{14.3}$$

With a constant Poisson's ratio, the shear modulus is expressed by

$$\mu = \frac{3k_b(1-2\nu)}{2(1+\nu)} \tag{14.4}$$

Since the bulk modulus is a function of the current elastic volumetric strain, and the elastic volumetric strain increases with the loading, the bulk modulus and the shear modulus increase with the loading. Therefore, in the elastic stage of the soil, the constitutive law of the soil is strongly nonlinear.

In the definition of the porous elasticity model in ANSYS, these material parameters, including swell index κ, elastic limit of tensile strength p_t^{el},

Poisson's ratio v, and initial void ratio e_0, are defined using the following format [4]

```
TB, PELAS,,,,POISSON
TBDATA, 1,κ, p_t^el, v, e0.
```

Furthermore, if the initial stress p0 exists in the soil, it can be specified as [4]

```
INISTATE, set, dtyp, stre
INISTATE, defi,all,,,,-p0,-p0,-p0   ! negative means
                                      compression
```

14.2.2 Plastic Component

In ANSYS, the plastic part of the Cam Clay model was developed based on the Extended Cam Clay model with the following yield function (see Figure 14.4) [4]:

$$f_{cc} = \frac{1}{\beta^2}\left(\frac{p}{a_h} - 1\right)^2 + \left(\frac{t}{M_c a_h}\right)^2 - 1 \qquad (14.5)$$

in which β is used to modify the shape of the yield surface, including β_{dry} for the dry side and β_{wet} for the wet side. a_h is the parameter that controls the size of the yield surface. M_c is the slope of the critical state line. Parameter t is defined by

$$t = \frac{\sigma_e}{g} = \frac{\sigma_e^3(1+K_s)+13.5(1-K_s)J_3}{2K_s\sigma_e^2} \qquad (14.6)$$

In the above equation, σ_e is the von Mises effective stress, J_3 is the third principal invariants of the stress tensor, and K_s modifies the shape of the yield surface (see Figure 14.5). The relation between the void ratio and the pressure during plastic deformation is given as

$$de = -\lambda_F d(\ln p) \qquad (14.7)$$

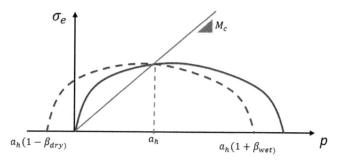

FIGURE 14.4
Yield function of the Extended Cam Clay model [4].

Cam Clay Model

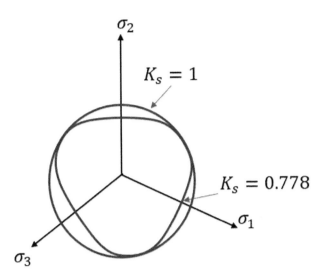

FIGURE 14.5
Various yield functions with different K_s.

Thus, seven parameters are defined in the Cam Clay model in ANSYS [4]:

```
TB, SOIL,1,,,CAMCLAY
TBDATA,1, λ_F, Mc, ah0, ah,min, β_dry
TBDATA,6, β_wet, Ks
```

14.3 Simulation of a Tower on the Ground by Cam Clay Model

Subsidence occurs due to the deformation of the soils when the building is constructed. Since subsidence is always associated with the stability of the building, it is very important in the structural design [3]. In this study, the Modified Cam Clay model was applied to study the subsidence of the tower built, which provides an example for a similar structural analysis.

14.3.1 Finite Element Model

Towers are generally conical. Thus, one cross-section of the model was created in ANSYS190 (see Figure 14.6), which was meshed by PLANE182 with axisymmetrical keyoption (keyopt(3)=1).

14.3.2 Material Properties

The tower was modeled with a linear elastic material with Young's modulus 2e10 Pa and density 5000kg/m^3. The soil was simulated with the porous

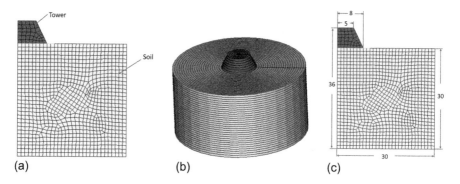

FIGURE 14.6
Finite element model of a tower on the ground. (a) 2D axisymmetrical model. (b) 3D model. (c) Geometry (all dimensions in meter).

elasticity and Cam Clay model. The corresponding APDL commands to define the soil material parameters are listed as

```
TB,PELAS,1,,,POISSON
TBDATA,1,0.026,5.0e3,0.28, 0.889
TB,SOIL,1,,,CAMCLAY
TBDATA,1,0.1174,1,103e3, 103e3,1
TBDATA,6,1,1
```

14.3.3 Contact Definition

The contact between the tower and the ground was defined as standard contact. Because the tower is much stiffer than the ground, the contact areas connected with the tower and the ground were modeled with TARGE169 and CONTA172, respectively (see Figure 14.7).

14.3.4 Loadings and Boundary Conditions

The whole model was loaded with self-weight using command ACEL, and the ground was constrained on the boundary (see Figure 14.8).

14.3.5 Results

The deformation of the whole model is illustrated in Figure 14.9, which indicates that the vertical displacement 2.69m occurs. The von Mises stresses and the plastic strains of the soil are plotted in Figures 14.10 and 14.11, respectively. Obviously, the stresses increase with the depth of the soil because of the weight. The distribution of the plastic strains looks like that of the stresses, except that the maximum plastic strain 0.075 occurs at the contact edge between the ground and the tower.

Cam Clay Model

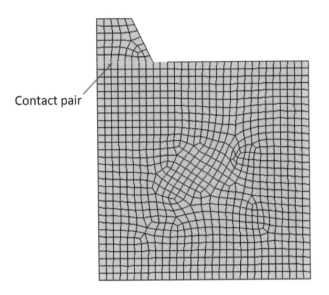

FIGURE 14.7
Contact in the model.

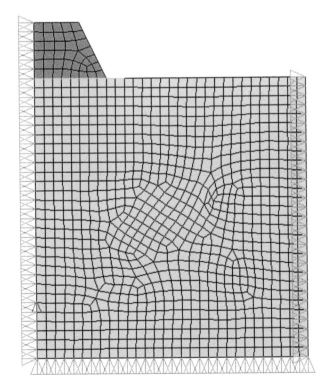

FIGURE 14.8
Boundary conditions and loadings.

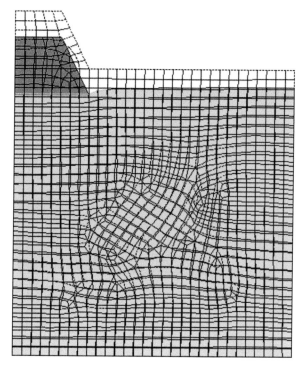

FIGURE 14.9
Deformation (Dmax = 2.69 m).

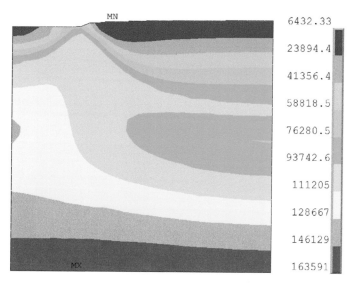

FIGURE 14.10
vM stresses of the soil (Pa).

Cam Clay Model

The history of the vertical displacement of the tower with the loading is presented in Figure 14.12, which shows that, at the beginning stage, the vertical displacement increases quickly with the increase of the loading. After 10% of the loading was loaded, the vertical displacement increased much more slowly than that at the beginning stage.

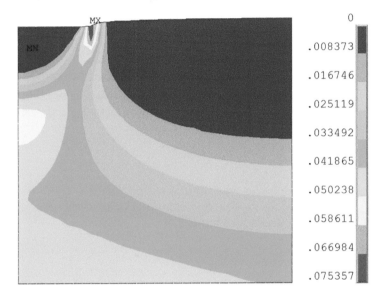

FIGURE 14.11
vM plastic strains of the soil.

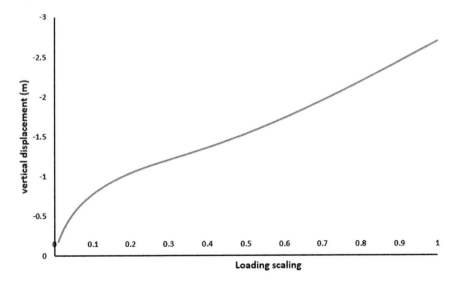

FIGURE 14.12
Vertical displacement with loading scaling.

14.3.6 Discussion

A 2D axisymmetrical finite element model was constructed to simulate the subsidence of the tower built using the porous elasticity and Cam Clay model. The large vertical displacement of the tower occurs after the whole model is loaded with self-weight, which confirms the subsidence of the tower.

The loading history of the vertical displacement of the tower reveals the features of the soil model. The stiffness of the soil at the elasticity increases with the pressure that is linked to the loading. At the beginning, the stiffness of the soil is very small. Therefore, the vertical displacement changes rapidly. With the increase of the loading, the pressure increases as well, which escalates the stiffness of the soil. Thus, the vertical displacement changes slow gradually. That is why the vertical displacement changes with the loading differ before and after 10% of the loading.

14.3.7 Summary

The subsidence of the tower was simulated in ANSYS190 using the porous elasticity and Cam Clay model. The computational results are consistent with the features of the material models.

References

1. Roscoe, K.H., Schofield, A.N., Wroth, C.P., On the yielding of soils, *Geotechnique*, Vol. 8, 1958, pp. 22–53.
2. Roscoe, K.H., Burland, J.B., On the generalised stress-strain behaviour of 'wet' clay, *Engineering Plasticity*, Cambridge University Press, 1968, pp. 535–609.
3. Helwany, S., *Applied Soil Mechanics with ABAQUS Application*, John Wiley & Sons, Inc., 2007.
4. ANSYS190 Help Documentation in the help page of product ANSYS190.

15

Drucker–Prager Model

In addition to the Cam Clay model, the Drucker–Prager model is another widely used model in geomechanics. Chapter 15 briefly introduces the Drucker–Prager model and its definition in ANSYS in Section 15.1; it describes its application for analysis of a soil–arch interaction in Section 15.2.

15.1 Introduction of Drucker–Prager Model

The Drucker–Prager model is a pressure-dependent model that is widely applied for the study of granular (frictional) materials like soils, rock, and concrete. In ANSYS, some material models related to the Drucker–Prager model are available, including the classic Drucker–Prager model, Extended Drucker–Prager (EDP) model, Extended Drucker–Prager Cap model, and Drucker–Prager Concrete model. These models all follow the same yield function [1],

$$f(\sigma, \sigma_y) = \sigma_e + \alpha \frac{1}{3} tr(\sigma) - \sigma_y \qquad (15.1)$$

where
 σ_e is equivalent stress,
 σ_y is uniaxial yield stress, and
 α is pressure sensitivity.

α is a function of internal frictional angle ϕ,

$$\alpha = \frac{6\sin\phi}{\sqrt{3}(3-\sin\phi)} \qquad (15.2)$$

and σ_y is expressed in terms of a function of ϕ and cohesion value c

$$\sigma_y = \frac{6c\cos\phi}{\sqrt{3}(3-\sin\phi)} \qquad (15.3)$$

The yield function of the Drucker–Prager model in π plane is circular (see Figure 15.1). Because the yield surface of the Drucker–Prager model is smooth, it has no convergence difficulty in large engineering problems.

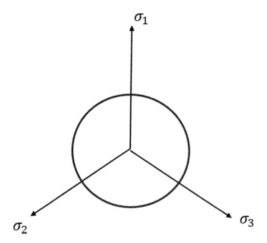

FIGURE 15.1
Yield surface of Drucker–Prager model in π plane.

Thus, the Drucker–Prager model has been widely used for the simulation of rocks and concretes.

In ANSYS, the Extended Drucker–Prager (EDP) model is specified by

```
! define the yield surface
TB, EDP,,,,LYFUN
TBDATA,1, α, σy
! define the plastic flow potential
TB,EDP,,,,LFPOT
TBDATA,1, ᾱ        ! ᾱ is pressure sensitivity
```

The definition of other Drucker–Prager models can also be found in the ANSYS190 Help Documentation [1].

15.2 Study of a Soil–Arch Interaction

A strip footing on soil was analyzed by Zienkiewicz [2]. In this study, the problem was analyzed using the Drucker–Prager model in ANSYS190. The model's convergence pattern was compared with that of the Mohr–Coulomb model.

15.2.1 Finite Element Model

A simple soil–arch interaction model is illustrated in Figure 15.2 [3]. The arch has a 3 m span, 750 mm rise, and 215 mm thickness. The arch and soil

Drucker–Prager Model

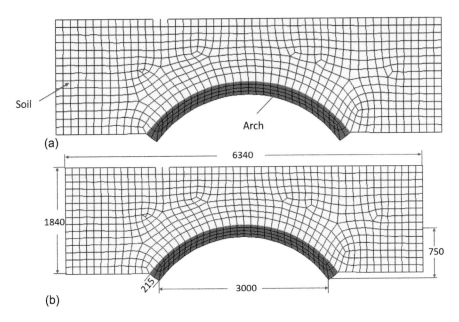

FIGURE 15.2
Finite element model of a soil–arch interaction. (a) Finite element model (b) Geometry (all dimensions in mm).

were meshed by Plane182 with a plane strain state. The contact between the soil and arch barrel interface was assumed standard contact and meshed by CONTA173 and TARGE170. The whole model consists of 818 elements and 856 nodes.

15.2.2 Material Properties

The arch barrel top was regarded as elastic with Young's modulus 16GPa and had a Poisson ratio 0.2. The soil was modeled as Drucker–Prager material [3]:

```
! set nonlinear parameters for intact rock
beta=1.05       ! Drucker-Prager parameter (slope of cone)
sigma_y=0.0224  ! Drucker-Prager parameter (strength)
TB,EDP,3,1,2,LYFUN
tbdata,1,beta,sigma_y

TB,EDP,3,1,2,LFPOT
tbdata,1,beta
```

The coefficient of friction between the soil and arch barrel was assumed to be 0.7.

15.2.3 Boundary Conditions and Loadings

The whole soil area was constrained in the normal directions of the boundary (see Figure 15.3), and the two ends of the arch were constrained with all degrees of freedom.

A pressure 0.3MPa was applied to a part of the top edge of the soil.

15.2.4 Results

The final deformation, von Mises stresses, and the plastic strains of the whole model are presented in Figures 15.4 through 15.6, respectively. The maximum

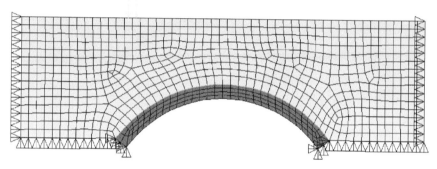

FIGURE 15.3
Boundary conditions and loadings.

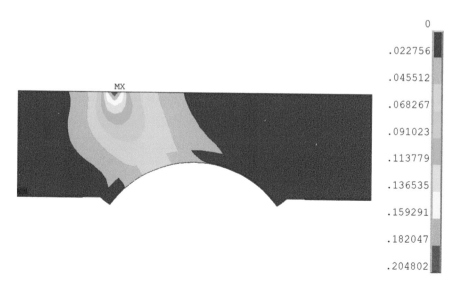

FIGURE 15.4
Deformation of the model (mm).

Drucker–Prager Model

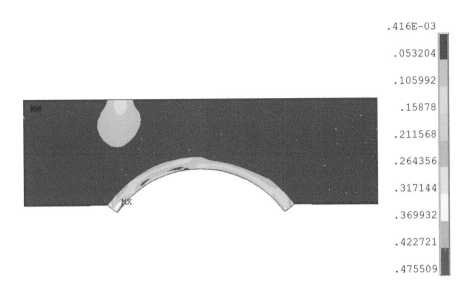

FIGURE 15.5
vM stresses of the model (MPa).

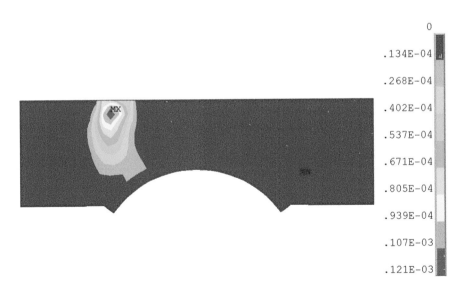

FIGURE 15.6
vM plastic strains of the model.

displacement 0.2 mm occurs at the loading area, and the maximum von Mises stresses 0.48 MPa appear within the arch because the arch is 16 times stiffer than the soil. The maximum plastic strain 1.2e-4 is just a little below the loading area.

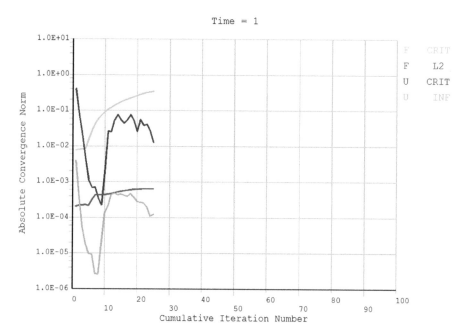

FIGURE 15.7
Convergence pattern of the model.

Figure 15.7 plots the convergence pattern of the model during the solution, which indicates it converges quickly and finishes all computations in less than 30 iterations.

15.2.5 Discussion

A 2D finite element model of a soil–arch interaction was developed, in which the soil was modeled by the Drucker–Prager model. The stress state and plastic strain distribution were obtained after the solution was complete.

Compared to the Mohr–Coulomb model in Chapter 16, the Drucker–Prager model converges quickly because the yield surface of the Drucker–Prager model is smooth, and the standard Newton–Raphson method is applied to solve the nonlinear equations. On the other hand, the yield surface of the Mohr–Coulomb model has sharp angles. Therefore, the cutting-plane algorithm is used to solve the nonlinear equations in the Mohr–Coulomb models, which requires many iterations to complete.

15.2.6 Summary

The soil–arch interaction was simulated in ANSYS190 by the Drucker–Prager model. It converges much more quickly than the Mohr–Coulomb model.

References

1. ANSYS190 Help Documentation in the help page of product ANSYS190.
2. Zienkiewicz, O.C., Humpheson, C., Lewis, R.W., Associated and non-associated viscoplasticity and plasticity in soil mechanics, *Geotechnique*, Vol. 25, 1975, pp. 671–689.
3. Wang, J., Melbourne, C., Finite element analyses of soil-structure interaction in masonry arch bridges, Arch's 07 – 5th International Conference on Arch Bridge, Madeira, September 2007, pp. 515–523.

16

Mohr–Coulomb Model

Chapter 16 mainly discusses the Mohr–Coulomb model. The introduction of the Mohr–Coulomb model, definition of the Mohr–Coulomb model in ANSYS, and application of the Mohr–Coulomb model for slope stability analysis are presented in Sections 16.1 to 16.3, respectively.

16.1 Introduction of Mohr–Coulomb Model

When the soil is under triaxial compression (see Figure 16.1), the soil fails along a failure plane at an angle θ with respect to σ_1 by shearing. Thus, the Mohr–Coulomb model was presented to define the failure when the shear stress in the material reaches the following criterion (see Figure 16.2):

$$\tau = c - \sigma \tan \phi \tag{16.1}$$

in which
τ – shear stress;
c – cohesion;
σ – normal stress; and
ϕ – inner friction angle.

The relationship between the angle θ and ϕ is illustrated in Figure 16.3:

$$\theta = 45° + \frac{\phi}{2} \tag{16.2}$$

Also, Figure 16.3 shows that

$$s = \frac{\sigma_1 - \sigma_3}{2} \tag{16.3}$$

$$\tau = s \cos \phi \tag{16.4}$$

$$\sigma = \sigma_m + s \sin \phi \tag{16.5}$$

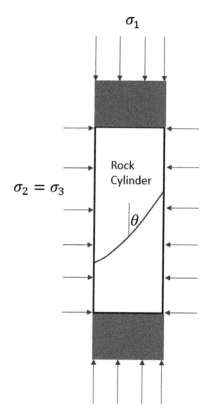

FIGURE 16.1
Soil under triaxial compression.

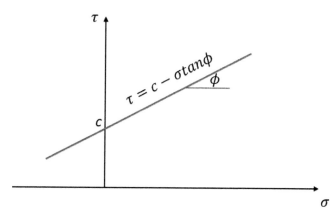

FIGURE 16.2
$\tau - \sigma$ of Mohr–Coulomb model.

Mohr–Coulomb Model

Substituting the above three equations into equation (16.1) results in the following:

$$s + \sigma_m \sin\phi - c \cos\phi = 0 \qquad (16.6)$$

In the above equation, s is maximum shear stress, and $\sigma_m = \dfrac{\sigma_1 + \sigma_3}{2}$ is the average of maximum stress and minimum stress. The yield surface of the Mohr–Coulomb model in π plane is a conical prism (see Figure 16.4).

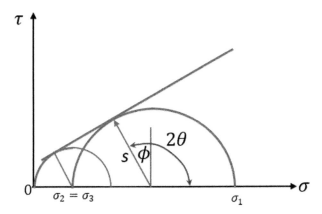

FIGURE 16.3
Mohr–Coulomb failure envelope.

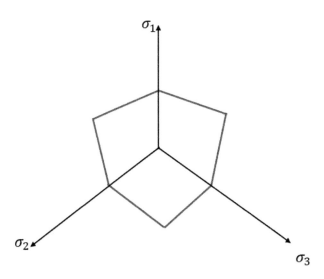

FIGURE 16.4
Yield surface of Mohr–Coulomb model in π plane.

16.2 Mohr–Coulomb Model in ANSYS

Under the triaxial compression, Figure 16.5a shows the true stress–strain curve of the soil, which can be approximated by a bilinear curve (see Figure 16.5b). Therefore, the plastic theory is applied for the Mohr–Coulomb model.

At the yield surface,

$$f(\sigma_1, \sigma_2, \sigma_3) = 0 \tag{16.7}$$

For flow rule of plastic strain

$$\dot{\varepsilon}_i^p = \lambda \frac{\partial g}{\partial \sigma_i} \tag{16.8}$$

If $g = f$, it is classical associated plasticity. Otherwise, it is general non-associated plasticity.

In ANSYS, the Mohr–Coulomb yield surface is defined as [1]

$$f_{MC}(\sigma) = \frac{\sigma_e}{\sqrt{3}}\left(\cos\vartheta - \frac{\sin\vartheta \sin\phi}{\sqrt{3}}\right) - \sigma \sin\phi - c\cos\phi \tag{16.9}$$

where

$$\sigma_e = \sqrt{3 J_2} \tag{16.10}$$

$$\sin(3\vartheta) = -\frac{3\sqrt{3}}{2} \frac{J_3}{\sqrt{J_2^3}} \tag{16.11}$$

In tension, the yield surface of the Mohr–Coulomb model is specified as [1]

$$f_R(\sigma) = \frac{2}{3}\sigma_e \sin(\vartheta + 120°) - \sigma - T \tag{16.12}$$

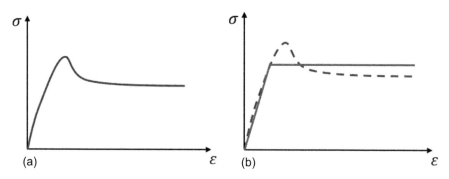

FIGURE 16.5
Real stress–strain curve of soil and its approximation. (a) Real soil. (b) Bilinear approximation.

where T is the tensile strength.

The Mohr–Coulomb flow potential is described by [1]

$$g_{MCt} = \frac{\sigma_e}{\sqrt{3}}\left(\cos\vartheta - \frac{\sin\vartheta\sin\psi}{\sqrt{3}}\right) - \sigma\sin\psi \qquad (16.13)$$

where ψ is the dilatancy angle.

Therefore, the Mohr–Coulomb model needs to input the following material parameters in ANSYS [1]:

```
TB,    MC,  ,  ,  ,  BASE
TBDATA, 1, ϕ, c, ψ, ϕ', , c'
TB,    MC,  ,  ,  ,  RCUT
TBDATA, 1, T, T'
```

ϕ', c', and T' are the residual inner friction angle, residual cohesion, and residual tensile strength, respectively.

ANSYS also provides other optional parameters to define the softening of the Mohr–Coulomb model. Please check the ANSYS help documentation for details.

The Mohr–Coulomb model is widely used to represent aggregate materials such as soil, rock, and concrete that start plastic deformation when the shear stress reaches a critical value. The Mohr–Coulomb model is also applied to study the stability of a slope.

16.3 Study of Slope Stability

Landslides are often reported in the news. The slope stability associated with the landslide closely refers to the inclined soil's ability to withstand the slope movement. Some methods, such as kinematic analysis and limit equilibrium analysis, have been developed to study slope stability [2, 3]. Due to advancements in technology in the past two decades, the slope stability analysis has been investigated using the finite element method [4–6]. In this study, the slope stability of a slope was modeled by the Mohr–Coulomb model and studied using the finite element strength reduction method.

16.3.1 Finite Element Model

Half of a slope was developed in ANSYS190 (see Figure 16.6) because of its symmetry. The slope was meshed by PLANE182 (plane strain) with 1,050 elements and 1,116 nodes.

16.3.2 Material Properties

The slope was modeled with the Mohr–Coulomb material model with different trial safety factors F_t listed in Table 16.1. The material parameters of the Mohr–Coulomb model with different F_t are calculated by

$$c_t = \frac{c}{F_t} \tag{16.14}$$

$$\phi_t = \arctan\left(\frac{\tan \phi}{F_t}\right) \tag{16.15}$$

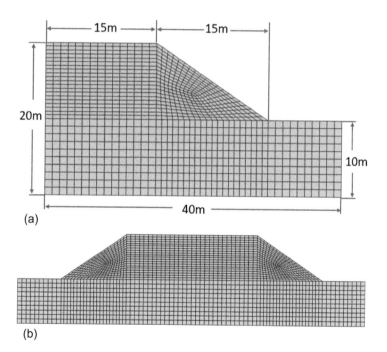

FIGURE 16.6
Finite element model of a slope. (a) Half a model. (b) Full model.

TABLE 16.1
Material parameters of Mohr–Coulomb material model

Ft	$\varphi(°)$	C (kPa)
1.00	25	15
1.40	18.4	10.71
1.60	16.25	9.38
1.63	15.96	9.20

The Mohr–Coulomb model was defined by the following:

```
TB,MC,1,,,BASE
TBDATA,1,17.2,10e3,0, 17.2,10e3  !c=15kPa, fictional
angle 25°
TB,MC,1,,,RCUT
TBDATA,1,0.5e3,0.5e3
TB,MC,1,,,MSOL
TBDATA,1,2              ! elastic tangent
```

16.3.3 Loadings and Boundary Conditions

The whole model was applied with self-weight by command ACEL, and the boundary of the slope was constrained (see Figure 16.7).

16.3.4 Results

The equivalent plastic strains with different trial safety factors F_t are plotted in Figure 16.8. The figure clearly shows that when $F_t = 1.4$, the plastic zone starts at the corner of the slope. When F_t increases to 1.6, the plastic zone gradually moves up. When $F_t = 1.63$, the plastic zone approaches the top of the slope. The whole plastic zone then looks like a curve. Thus, $F_t = 1.63$ should be the safety factor of the slope stability.

The deformation and von Mises stresses of the slope when $F_t = 1.63$ are illustrated in Figures 16.9 and 16.10, respectively. Due to the loading of self-weight, the stresses of the slope increase with the increase of the depth and reach the maximum at the bottom with value 0.216 MPa. The maximum displacement 0.52 mm occurs at the top.

The typical convergence pattern of the Mohr–Coulomb material model is presented in Figure 16.11, which shows that it converges slowly and takes a lot of substeps to complete the entire computation.

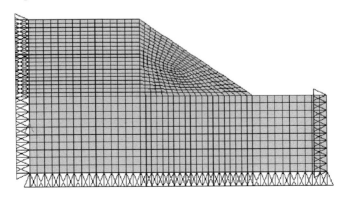

FIGURE 16.7
Loadings and boundary conditions.

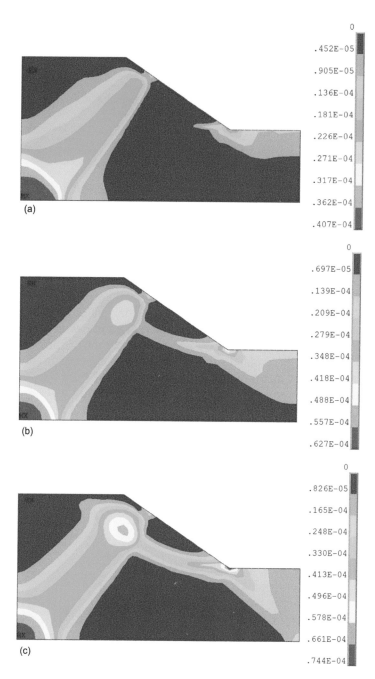

FIGURE 16.8
Equivalent plastic strains with different trial safety factors F_t. (a) $F_t=1.4$. (b) $F_t=1.6$. (c) $F_t=1.63$.

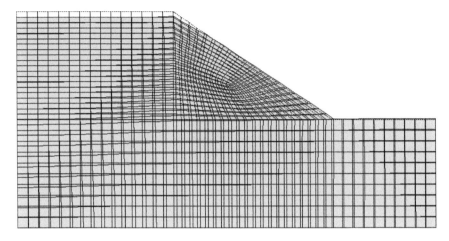

FIGURE 16.9
Deformation of the soil (Dmax = 0.52 m).

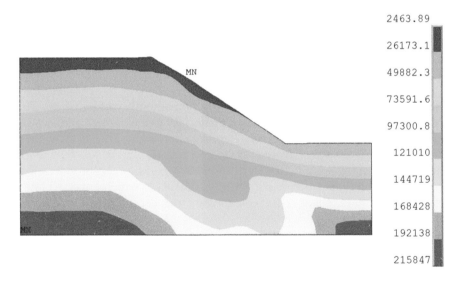

FIGURE 16.10
vM stresses of the soil (F_t =1.63) (Pa).

16.3.5 Discussion

A 2D plane strain finite element model was created to study the stability of a slope. After comparing the plastic zones obtained by different trial safety factors, the safety factor of the slope was defined as the obtained plastic zone extends from the bottom to the top of the slope.

In ANSYS, the Mohr–Coulomb and Jointed Rock models adopt the cutting-plane algorithm to solve the nonlinear equations because the Mohr–Coulomb

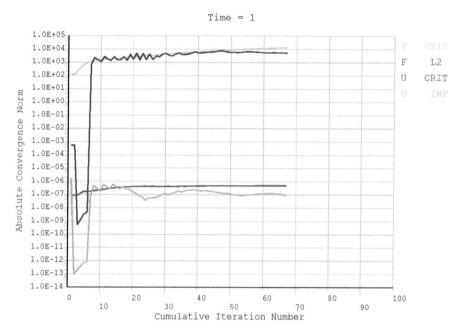

FIGURE 16.11
Convergence pattern of Mohr–Coulomb model.

and Jointed Rock material models have a non-smooth yield surface. This cutting-plane algorithm has an elastic material tangent, which reduces the convergence speed of the Newton–Raphson iterations. That is why these material models always converge slowly and require a lot of substeps to finish the solution. In ANSYS, the cutting-plane algorithm can be selected in the material solution option using TB, MC,,,,MSOL, in which options 1 and 2 refer to consistent tangent and elastic tangent, respectively.

16.3.6 Summary

The slope stability was studied with the Mohr–Coulomb model. The safety factor of the slope was defined by comparing the plastic zones obtained by different trial safety factors.

References

1. ANSYS190 Help Documentation in the help page of product ANSYS190.
2. Eberhardt, E., *Rock Slope Stability Analysis – Utilization of Advanced Numerical Techniques* (PDF), Vancouver, Canada: Earth and Ocean Sciences, University of British Columbia, 2003.

3. Bishop, A.W., The use of the slip circle in the stability analysis of slopes, *Geotechnique*, Vol. 5, 1955, pp. 7–17.
4. Donald, I., Chen, Z.Y., Slope stability analysis by the upper bound approach: fundamentals and methods, *Canada Geotechnical Journal*, Vol. 34, 1997, pp. 853–862.
5. Griffiths, D.V., Lane, P.A., Slope stability analysis by finite elements. *Geotechnique*, Vol. 49, 1999, pp. 387–403.
6. Xu, X., Dai, Z., Numerical implementation of a modified Mohr-Coulomb model and its application, *Journal of Modern Transportation*, Vol. 25, 2017, pp. 40–51.

17
Jointed Rock Model

Geologic and aggregate materials are inhomogeneous and contain weak planes in the materials, such as stratification planes. This kind of plane is named a jointed rock, which is anisotropic and non-linear, and has lower stiffness, higher permeability, and less strength than the intact rock. Various methods have been proposed to take the influence of discontinuities into account on the mechanical behavior of jointed rock [1–6]. Recently, ANSYS also released the Jointed Rock model for geomechanics simulation, which is introduced as follows.

17.1 Jointed Rock Model

The yield surface on a joint j is expressed in the form of the Mohr–Coulomb model as [7]

$$f_j = \tau_j - \sigma_j \tan\phi_j - c_j \tag{17.1}$$

where
τ_j – shear stress on joint j,
σ_j – normal stress on joint j,
ϕ_j – frictional angle at joint j,
c_j – cohesion at joint j.

The flow potential for the joint is [7]

$$Q_j = \tau_j - \sigma_j \tan\psi_j \tag{17.2}$$

where ψ_j is the dilatancy angle.

The tension-cutoff yield surface is defined as [7]

$$f_{Tj} = \sigma_j - T_j \tag{17.3}$$

where T_j is the tensile strength of joint j.

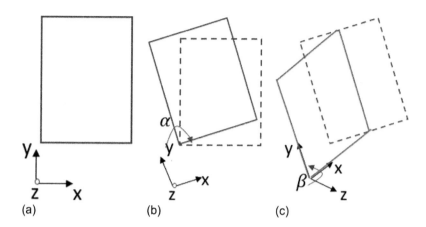

FIGURE 17.1
Determination of orientation of a joint [7]. (a) Initial position. (b) Rotate angle α clockwise around axis Z. (c) Rotate angle β around axis Y counterclockwise.

The residual yield surface and residual tensile strength yield surface follow equations (17.1) and (17.3), respectively, with different coefficients.

The orientation of the joint j separation plane is described by two angles (see Figure 17.1) [7]:

(1) rotate angle α around Z axis clockwise and (2) rotate angle β around Y axis counterclockwise.

17.2 Definition of the Jointed Rock Model in ANSYS

The definition of the Jointed Rock model in ANSYS requires a clarification of both the base material and the joints.

17.2.1 Defining the Base Material

The definition of the base material of the Jointed Rock model is the same as the one for the Mohr–Coulomb model [7]:

```
TB, JROCK,,,, BASE
TBDATA, 1, ϕ,c_j,ψ_j,ϕ',c'_j
TB, JROCK,,,, RCUT
TBDATA, 1, T, T'
TB, JROCK,,,, RSC
TBDATA, 1, R_SC
```

Jointed Rock Model

17.2.2 Defining the Joints

As many as four joints can be defined in ANSYS. Each joint demands yield stress and flow-potential parameters at the joint [7]:

```
TB, JROCK,,,,FPLANE
TBDATA, 1, ϕⱼ, cⱼ, ψⱼ, ϕ'ⱼ, c'ⱼ
TB, JROCK,,,,FTCUT
TBDATA,1, T,T'
```

Orientation of the joint is determined by the rotation angles α and β (see Figure 17.1) that are imported by the TB command:

```
TB, JROCK,,,,FORIE
TBDATA, 1, α, β
```

The above ANSYS commands can be repeated to define more joints.

17.3 Simulation of Tunnel Excavation

Tunnel construction requires excavation of the soil. Understanding the stress state of the soil during tunnel excavation is very important for the project design. In this study, the finite element method was applied to study the tunnel excavation using the Jointed Rock material model.

17.3.1 Finite Element Model

A tunnel with a diameter of D=1.5m was excavated from a 10m-width square [8]. Thus, the whole model was composed of the tunnel (purple in Figure 17.2) and the square without the tunnel (blue in Figure 17.2) and meshed by PLANE182 (plane strain) with 1,669 elements and 1,749 nodes.

17.3.2 Material Properties

The area around the tunnel was modeled with the Jointed Rock material model. The base material regarded as the Mohr–Coulomb material was defined by

```
TB,JROCK,1,,,BASE
TBDATA,1,21.2,1e6,10,21.2,1e6
TB,JROCK,1,,,RCUT
TBDATA,1,0.1e6,0.1e6

TB,JROCK,1,,,RSC
TBDATA,1,0
```

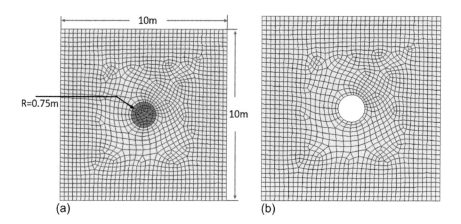

FIGURE 17.2
Finite element model of tunnel excavation. (a) Before excavation. (b) After excavation.

The Jointed Rock, including its material model and orientation of the joint failure plane, was specified by

```
TB,JROCK,1,,,FPLANE
TBDATA,1,15.8,0.1e6,0,15.8,0.1e6

TB,JROCK,1,,,FTCUT
TBDATA,1,0.05e6,0.05e6

TB,JROCK,1,,,FORIE
TBDATA,1,,-60
```

17.3.3 Loadings and Boundary Conditions

The boundary of the whole model was fixed (see Figure 17.3), and the initial stress 1e6MPa was applied on all the elements in both X and Y directions by

```
inistate,set,dtyp,stre
inistate,defi,all,,,,,-1e6,-1e6,,0,0,0
```

17.3.4 Solution

The solution was completed in two steps. The first step solved the full square. In the second step, the elements within the central tunnel part were deactivated using the EKILL command to simulate excavation.

17.3.5 Results

Figure 17.4 illustrates the final deformation of the model. The displacements mainly occur around the tunnel with maximum displacement 7.35 mm. The von Mises stresses of the model are plotted in Figures 17.5 and 17.6. In the

Jointed Rock Model

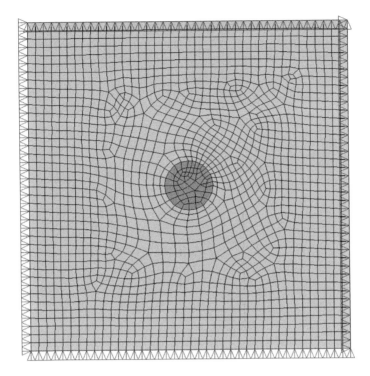

FIGURE 17.3
Boundary conditions.

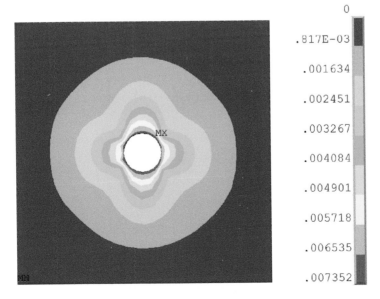

FIGURE 17.4
Final deformation (m).

144

FIGURE 17.5
Stresses of the soil at the end of the first step (Stress = 847950 Pa).

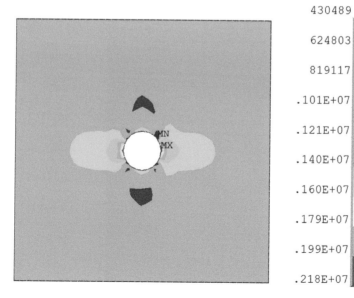

FIGURE 17.6
Final vM stresses of the soil (Pa).

Jointed Rock Model

first step, the stress distribution is uniform with 0.85MPa. The stress concentration occurs around the edge of the tunnel in the second step because of the excavation. The final plastic strain contour is presented in Figure 17.7, which clearly shows that the main plastic strains occur at four different directions. For comparison, the final von Mises stresses of the model with the Mohr–Coulomb material model are given in Figure 17.8, showing that the stresses distribute uniformly around the edge of the tunnel and that no plastic strain occurs.

17.3.6 Discussion

A 2D plane strain finite element model was developed to simulate the tunnel excavation using the Jointed Rock material model. After excavation, the stress concentration occurs at the edge of the tunnel, and the major plastic strains are in four different directions, which match the behavior of the Jointed Rock material model. The same model without specifying the failure plane shows a uniform stress distribution around the edge of the tunnel and no plastic strain. The different results state clearly that the stress concentration and the major plastic strains are due to the failure plane of the Jointed Rock material model.

The EKILL command was applied to deactivate the elements to simulate excavation. Element birth and death is very useful in modeling. For example, contact element birth and death is used in simulation of the contact between

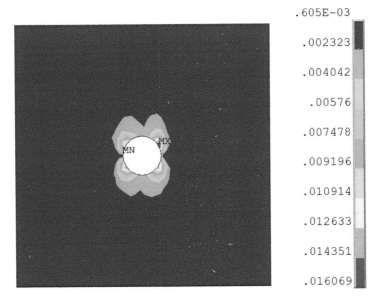

FIGURE 17.7
Final vM plastic strains of the soil.

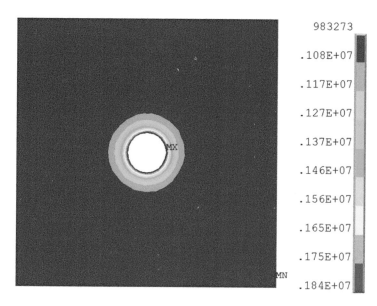

FIGURE 17.8
Final vM stresses of the soil with Mohr–Coulomb model only (Pa).

the stent and the vessel in angioplasty [9]. The same method is adopted in Chapter 21 to simulate the contact between the rod and the flange. In addition, initial stress was loaded to simulate the stress state at the beginning, which is another advanced finite element technology of ANSYS.

The whole computation was completed in two steps. The first step was elastic. Therefore, it was solved in one substep. Excavation in the second step made the stress state of the soil beyond the elastic stage and triggered material nonlinearity. Thus, it took 11 substeps to complete. Therefore, different substeps were selected for the two steps.

17.3.7 Summary

The tunnel excavation was simulated using the Jointed Rock material model. The computational results match the behavior of the material model.

References

1. Ramamurthy, T., A geo-engineering classification for rocks and rock masses, *International Journal of Rock Mechanics and Mining Sciences*, Vol. 41, 2004, pp. 89–101.
2. Jaeger, J., Shear failure of anisotropic rocks, *Geological Magazine*, Vol. 97, 1960, pp. 65–72.

3. Chen, W., Konietzky, H., Abbas, S.M., Numerical simulation of time-independent and-dependent fracturing in sandstone, *Engineering Geology*, Vol. 193, 2015, pp. 118–131.
4. Chakraborti, S., Konietzky, H., Walter, K., A comparative study of different approaches for factor of safety calculations by shear strength reduction technique for non-linear Hoek–Brown failure criterion, *Geotechnical and Geological Engineering*, Vol. 30, 2012, pp. 925–934.
5. Brzovic, A., Villaescusa, E., Rock mass characterization and assessment of block-forming geological discontinuities during caving of primary copper ore at the El Teniente mine, Chile, *International Journal of Rock Mechanics and Mining Sciences*, Vol. 44, 2007, pp. 565–583.
6. Bray, J., A study of jointed and fractured rock, *Rock Mechanics and Engineering Geology*, Vol. 5, 1967, pp. 117–136.
7. ANSYS190 Help Documentation in the help page of product ANSYS190.
8. Chang, L., Konietzky, H., Application of the Mohr-Coulomb yield criterion for rocks with multiple joint sets using Fast Lagrangian Analysis of Continua 2D (FLAC2D) software, *Energies*, Vol. 11, 2018, p. 614.
9. Yang, Z., *Finite Element Analysis for Biomedical Engineering Applications*, CRC Press, 2019.

18

Consolidation of Soils

Chapter 18 discusses the consolidation of soils. Section 18.1 introduces the consolidation of soils, Section 18.2 presents the modeling of porous media in ANSYS, and Section 18.3 focuses on the simulation of the consolidation.

18.1 Consolidation of Soils

Soil primarily consists of a solid and water. When the soil is under compression, the volumetric stiffness of water is much higher than that of the solid; therefore, the water absorbs the pressure immediately and flows away from the region of high pressure. Thus, the solid gradually takes over the loading and shrinks in volume. That is the process of soil consolidation.

When a building is erected over a layer of soil with low permeability and low stiffness, a large settlement of the building may occur over many years due to the consolidation. Also, soil consolidation significantly influences construction projects such as tunnel excavation and the construction of embankments. Therefore, it is necessary to conduct the numerical simulation of soil consolidation. ANSYS provides strong tools in this field, which are introduced in the following section.

18.2 Modeling Porous Media in ANSYS

Porous media are regarded as two-phase materials governed by an extended Biot's consolidation theory [1]. The fluid flow is described by Darcy's Law as

$$q = -k\nabla p \qquad (18.1)$$

where
  p – pore pressure,
  k – permeability, and
  ∇ – gradient operator.

The solid phase follows the equation

$$\nabla \cdot (\sigma'' - \alpha p I) + f = 0 \qquad (18.2)$$

where
f – body force,
α – Biot coefficient,
σ'' – Biot effective stress, and
I – second-order identity tensor.

Based on the equations (18.1) and (18.2), ANSYS developed Coupled Pore–Pressure–Thermal (CPT) elements such as CPT212 (2D 4-node), CPT213 (2D 8-node), CPT215 (3D 8-node), CPT216 (3D 20-node), and CPT217 (3D 10-node) to simulate the porous media. The material properties of porous media are defined by [2]

```
TB, PM, ,, , PERM
TBDATA, 1, 1e-8      ! define solid permeability k
TB, PM, ,, , BIOT    ! define Biot coefficient
TBDATA, 1, 0.5
```

18.3 Simulation of Consolidation of Three-Well Zone

Primary consolidation is the settlement process in which a vertical stress changes due to an applied load that causes water to be squeezed out of the soil. The process is complete when all the water in the soil is squeezed out. In this study, the consolidation was simulated in ANSYS190 using CPT elements.

18.3.1 Finite Element Model

The whole area is composed of three rock zones and two fault zones (see Figure 18.1). Three wells with radius 0.2 m are distributed uniformly within the middle soil zone. The top and bottom rock zones, as well as the fault zones, were modeled with the elastic material and meshed by PLANE182 in the plane strain state. The central soil zone was regarded as porous media and meshed with CPT212. The entire model consists of 3,995 elements and 4,151 nodes (see Figure 18.2).

18.3.2 Material Properties

The top and bottom rock zones were assumed linear elastic with Young's modulus 65GPa and Poisson's ratio 0.25. The middle soil zone was modeled as porous media with the material properties defined as follows:

Consolidation of Soils

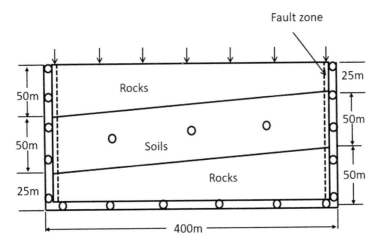

FIGURE 18.1
Consolidation of three-well zone.

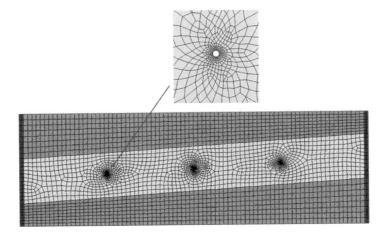

FIGURE 18.2
Finite element model of three-well zone.

```
fpx=1.0d-10
one=1.0

tb,pm,2,,,perm
tbdata,1,fpx,fpx,fpx
TB, PM, ,, , BIOT         ! define Biot coefficient
TBDATA, 1, one
```

The fault zone with Young's modulus 1GPa is much softer than the rock zones.

18.3.3 Boundary Conditions and Loadings

The whole area was constrained in the normal directions of the boundary (see Figure 18.3).

The total computation consists of two steps. In the first step, pressures 100 MPa and 50 MPa were loaded on the top surface and the well wall, respectively. Because the well walls had no flow, no pore pressure was defined there in the first step. However, the pore pressure of the well wall surfaces dropped to zero in the second step.

18.3.4 Solutions

The pressure loading in the first step was ramped (KBC, 0), and the loading in the second step was stepped (KBC, 1). Time in the first and second steps was set as one second and three hours, respectively.

18.3.5 Results

Figure 18.4 shows the final deformation of the whole model. The displacement decreases with the depth, and the maximum deformation 0.88 m occurs at the top of the fault zone because the fault zone is much softer than the rock zones. The final von Mises stresses are plotted in Figure 18.5. The stress is approximately uniformly distributed except for the top interface between the rock and fault zone. This is because at the final time, the water in the rock had almost completely been expelled into the well, and the rock zone had become purely solid. Under the pressure loading, the stresses of the rock should be close within the rock. The pore pressures of the model at the end of the first and second steps are presented in Figures 18.6 and 18.7,

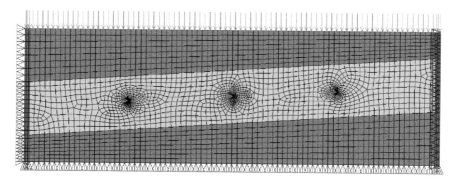

FIGURE 18.3
Boundary conditions and loadings.

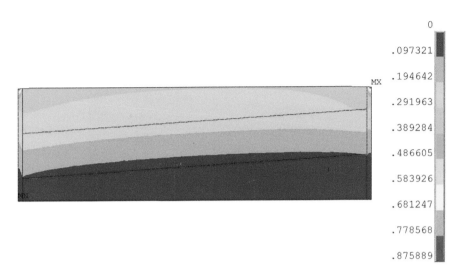

FIGURE 18.4
Final deformation of the model (m).

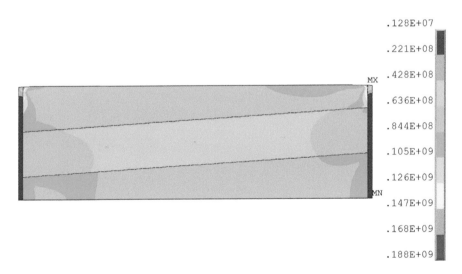

FIGURE 18.5
Final vM stresses of the model (Pa).

respectively, which clearly indicate that after three hours, the pore pressures dropped from about 98MPa to 1.0MPa. That is confirmed by the time history of the two nodes within the rock zone (see Figure 18.8). Node A is closer to the well than Node B. That is why the pore pressure of Node A drops faster than that of Node B at the beginning. After 6000s, both are nearly the same.

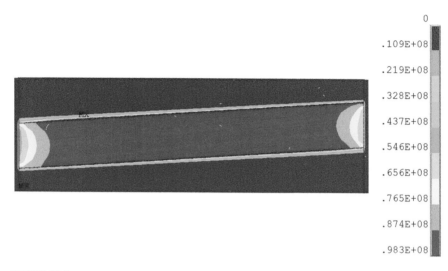

FIGURE 18.6
Pore pressures of the model at the end of the first step (Pa).

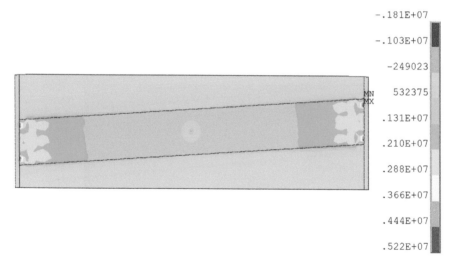

FIGURE 18.7
Final pore pressures of the model (Pa).

18.3.6 Discussion

A 2D plane strain finite element model was created to simulate the consolidation of the rocks using the pore media material model. After three hours, the water was expelled into the well, and the pore pressure within the rock dropped about 99%. This model confirms that the CPT elements can be used to simulate soil consolidation.

Consolidation of Soils

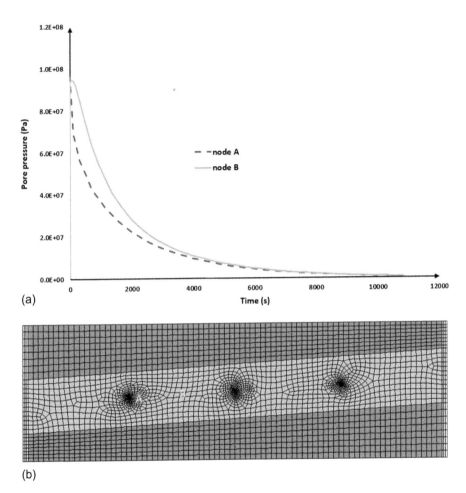

FIGURE 18.8
Time history of pore pressures of Nodes A and B. (a) Time history of pore pressures. (b) Locations of Nodes A and B.

Compared to SOLID elements such as SOLID185, the CPT elements have one extra DOF, pore pressure. Zero pore pressure was applied to define the water flow at the boundary. If no pore pressure is defined in the boundary, then the boundary has no water flow. Negative pore pressure refers to the area with suction effect.

18.3.7 Summary

The consolidation of the rocks was studied using CPT elements in ANSYS. After a long time, the pore pressure dropped to near zero because the water had been expelled into the well.

References

1. Biot, M.A., General theory of three-dimensional consolidation, *Journal of Applied Physics*, Vol. 12, 1941, pp. 155–164.
2. ANSYS190 Help Documentation in the help page of product ANSYS190.

Part IV

Modern Materials

In addition to the traditional materials discussed in the previous three parts, some modern materials, such as composites, shape-memory alloys, and functionally graded materials, have been developed and widely used in military aircraft, space, the automotive industry, and biomedical fields. The mechanical design of these materials demands an understanding of the stress and strain states of these materials; this can be achieved using the finite element method. Therefore, how to model these modern materials in finite element analysis becomes vital. That is the topic of Part IV.

Chapter 19 discusses composites, including how to model them in ANSYS, and provides two examples related to composite damage evolution and crack growth. Functionally graded materials are simulated in Chapter 20 using TBField technology in ANSYS. Chapter 21 focuses on shape-memory alloys and covers its application for orthodontic wire and a vacuum-tight shape-memory flange. Piezoelectric materials, as well as the simulation of a thin-film piezoelectric microaccelerometer, are presented in Chapter 22. The final chapter explains how to determine the Young's modulus of Fe nanoparticles from the experimental data in ANSYS.

19

Composite Materials

Composite materials have a long history, which can be traced back to the Israelites using straw to strengthen mud bricks. Modern composite materials were developed originally in the 1950s to address aerospace's demand for materials that had light weight, high strength, long fatigue life, and good thermal conductivity. Now, composite materials have been widely applied in military aircraft, civil aircraft, space, and the automotive industry.

Chapter 19 briefly introduces composite materials, describes modeling composite materials in ANSYS, and offers two applications of composite materials.

19.1 Introduction of Composite Materials

Composite material signifies that two or more materials are combined to form a third material. Unlike the combination on a microscopic scale, like the alloying of metals that is homogeneous macroscopically, the combination of materials in composite material, which can be identified by the naked eye, is non-homogenous on a macroscopic scale.

The composite materials are classified as

- Laminated composite materials consisting of layers of various materials (see Figure 19.1);
- Fibrous composite materials consisting of fibers in a matrix (see Figure 19.2);
- Particulate composite materials composed of particles in a matrix; and
- A combination of the above three types.

The well-designed composite materials usually exhibit better properties than any of their components or constituents. These properties improved by forming a composite material include but are not limited to the following [1]:

- Strength,
- Stiffness,

159

- Fatigue life,
- Corrosion resistance,
- Wear resistance,
- Thermal conductivity,
- Acoustical insulation, and
- Weight.

Composite materials can be designed to improve some of the above properties, but not all of them at the same time. Even some of the properties conflict with each other, such as thermal conductivity versus thermal insulation.

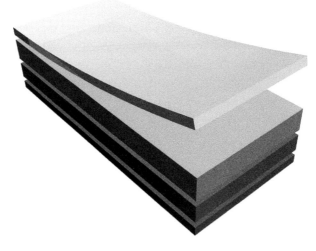

FIGURE 19.1
Illustration of the laminated composite (snehit©123rf.com).

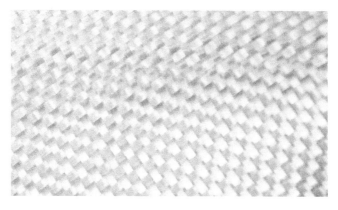

FIGURE 19.2
Fibrous composite material (prakasitlalao©123rf.com).

Composite Materials

Thus, the goal of designing composite materials is to enhance the specific properties required by the projects.

Composite materials also have some disadvantages [1]:

- Delamination – Composites always delaminate between layers;
- Damage inspection – Cracks and delamination in composites are usually internal and require special inspection techniques for detection;
- Complex fabrication – The fabrication of composites is always labor intensive and complex;
- High cost – Because the composite materials are new, they have a very high cost; and
- Composite to metal joining – The imbalance at joinery is due to the different thermal properties of the metal and composite and may lead to failure.

As composites play such an important role in modern industry, many features are available in ANSYS for modeling a composite. They are introduced in the next section.

19.2 Modeling Composite in ANSYS

19.2.1 Modeling Composite by Command SECTYPE

Many layers of fibrous composite materials are assembled to form a composite (see Figure 19.1). To define one layer, the following properties need to be specified:

(a) Layer thickness (TK),
(b) Material properties (MAT), and
(c) Layer orientation angle commands (THETA).

These parameters can be imported in ANSYS [2] as

```
sectype,1,shell
secdata,tk1,mat1,theta1  ! thickness, material ID, angle
    of 1st layer
secdata,tk2,mat2,theta2  ! thickness, material ID, angle
    of 2nd layer
            ...
secdata,tkn,matn,thetan  ! thickness, material ID, angle
    of nth layer
```

This defines composition, while the following explains that the composite damage can be characterized with damage-initiation criteria and damage-evolution law. The damage-initiation criteria are defined in ANSYS as

```
TB,DMGI,1,1,,FCRT ! FC for damage initiation
TBDATA, 1, C1 !Failure criteria type for the tensile
   fiber failure mode
TBDATA, 2, C2 !Failure criteria type for the compressive
   fiber failure mode
TBDATA, 3, C3 !Failure criteria type for the tensile
   matrix failure mode
TBDATA, 4, C4 !Failure criteria type for the compressive
   matrix failure mode
```

The damage-evolution law is specified in ANSYS by [2]

```
TB,DMGI,1,1,,MPDG ! FC for damage evolution
TBDATA, 1, C1 ! Tensile fiber stiffness reduction
TBDATA, 2, C2 ! Compressive fiber stiffness reduction
TBDATA, 3, C3 ! Tensile matrix stiffness reduction
TBDATA, 4, C4 ! Compressive matrix stiffness reduction
```

19.2.2 Modeling a Composite by Anisotropic Model

Unlike the most common engineering materials, composite materials are anisotropic. For a point of the body, anisotropic materials have material properties that differ with the directions. One special kind of anisotropic material is the orthotropic material with different material properties in three mutually perpendicular direction planes, which are defined in ANSYS using MP commands [2]:

```
MP, EX, ,E1   ! elastic modulus, element x direction
MP, EY, ,E2   ! elastic modulus, element y direction
MP, EZ, ,E3   ! elastic modulus, element z direction
MP, PRXY/NUXY, , v₁₂ ! major (minor) Poisson's ratio,
                       x-y plane
MP, PRYZ/NUYZ , , v₂₃ ! major (minor) Poisson's ratio,
                       y-z plane
MP, PRXZ/NUXZ , , v₁₃ ! major (minor) Poisson's ratio,
                       x-z plane
MP, GXY, , G₁₂   ! shear modulus, x-y plane
MP, GYZ, , G₂₃   ! shear modulus, y-z plane
MP, GXZ, , G₁₃   ! shear modulus, x-z plane
```

This definition as a simple way to specify that composite materials are widely used in the simulation of crack growth in composites.

19.3 Simulation of Composite Structure in Failure Test

The composite Structure Failure test is critical in determining the applied flight-load levels on a composite structure. It is performed until structural failure [3]. The aim of this study is to simulate the composite failure in ANSYS, which is helpful for understanding the failure mechanism of the composite structure.

19.3.1 Finite Element Model

The composite structure is composed of an adapter, load head, and conic part. The adapter connects the load head and conic part (see Figure 19.3). The conic part is a composite with a sandwich construction, in which the inner and outer composite faces are a $[(90/0)_4]$ layup. The finite element model of the composite structure was created in ANSYS190 (see Figure 19.4), in which the adapter and load head as well as the solid section of the conic part were meshed with SOLID185, and the composite section of the conic part was meshed with SHELL181. The whole model consists of 30,983 nodes and 130,395 elements.

19.3.2 Material Properties

The adapter and load head were assumed linear elastic with Young's modulus 180GPa and Poisson's ratio 0.3. Each composite ply of the conic part is 1.5 mm thick and composed of carbon/epoxy AS4-3501-6. The core of the sandwich composite is Rohacell 110WF with thickness 25 mm. Rohacell 110 WF was assumed as linear elastic with Young' modulus 26100psi and Poisson's ratio 0.286. AS4-3501-6 was modeled as orthotropic [3],

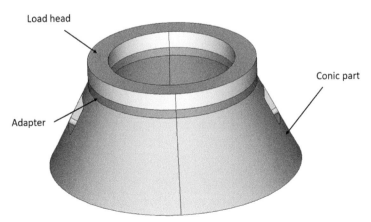

FIGURE 19.3
Structure of composite structure failure test.

FIGURE 19.4
Finite element model of the test.

```
!typical material properties for E-Glass/Epoxy (MPa)
e_x=18.4*6894
e_y=1.62*6894
e_z=e_y
pr_xy=0.1279
pr_yz=0.1331
pr_xz=pr_xy
g_xy=0.951*6894
g_yz=0.528*6894
g_xz=g_xy
MP,ex,1,e_x
MP,ey,1,e_y
MP,ez,1,e_z
MP,nuxy,1,pr_xy
MP,nuyz,1,pr_yz
MP,nuxz,1,pr_xz
MP,gxy,1,g_xy
MP,gyz,1,g_yz
MP,gxz,1,g_xz
```

The composite faces are a [(90/0)4], in which 0° and 90° are in the axial direction and the hoop direction, respectively. Therefore, a local coordinate system 11 is defined with X in the vertical direction as the element coordinate system of all shell elements (see Figure 19.5). The composite was defined in ANSYS by

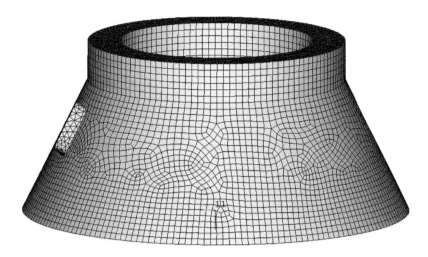

FIGURE 19.5
Local coordinate system 11 defined for element coordinate system.

```
sectype,1,shell
secdata,1.5,1,90      !# of section integration points
                         per layer
secdata,1.5,1,0
secdata,1.5,1,90
secdata,1.5,1,0
secdata,1.5,1,90      !# of section integration points
                         per layer
secdata,1.5,1,0
secdata,1.5,1,90
secdata,1.5,1,0
```

The parameters for damage evolution were specified in ANSYS by

```
TB,FCLI,1,1,,STRS ! material strengths
TBDATA,1,0.283*6894    !FAILURE STRESS, X DIRECTION TENSION
TBDATA,2,-0.215*6894   !FAILURE STRESS, X DIRECTION COMPRESSION
TBDATA,3,6.96e-3*6894  !FAILURE STRESS, Y DIRECTION TENSION
TBDATA,4,-2.90e-2*6894 !FAILURE STRESS, Y DIRECTION COMPRESSION
TBDATA,5,6.96e-3*6894  !FAILURE STRESS, Z DIRECTION TENSION
TBDATA,6,-2.90e-2*6894 !FAILURE STRESS, Z DIRECTION COMPRESSION
TBDATA,7,1.15e-2*6894  !FAILURE STRESS, XY DIRECTION TENSION
TBDATA,8,7.25e-3*6894  !FAILURE STRESS, YZ DIRECTION TENSION
TBDATA,9,1.15e-2*6894  !FAILURE STRESS, XZ DIRECTION TENSION
```

```
TB,DMGI,1,1,,FCRT ! FC for damage initiation
TBDATA,1,2,2,2,2 ! Max stress criteria for all four failure
                   modes
.
TB,DMGE,1,1,,MPDG ! Damage evolution with MPDG method
TBDATA,1,0.6 ! 60% fiber tension damage (40% ultimate
               strength)
TBDATA,2,0.6 ! 60% fiber compression damage (40% ultimate
               strength)
TBDATA,3,0.6 ! 60% matrix tension damage (40% ultimate
               strength)
TBDATA,4,0.6 ! 60% matrix compression damage (40% ultimate
               strength)
```

19.3.3 Boundary Conditions and Loadings

The surface of the load head was loaded with pressure 240 MPa. The bottom surface of the conic part was fixed with all degrees of freedom (see Figure 19.6).

19.3.4 Results

The final deformation and stresses of the model with damage were plotted in Figure 19.7. For comparison, Figure 19.8 presents the final deformation and

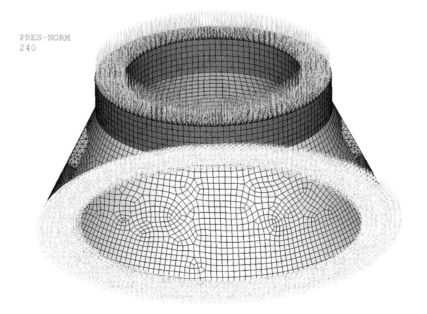

FIGURE 19.6
Boundary conditions and loading.

stresses of the model without damage. Both cases have very similar stress distributions because of the force loading. However, the model with damage has a much larger deformation than that without damage. It was confirmed by the displacement history of one node located on the top surface (see Figure 19.9). It clearly shows that at the beginning, in the elastic stage, both models have an identical path. After the damage initiates around loading 180 MPa, the displacement in the model with damage starts to shift from

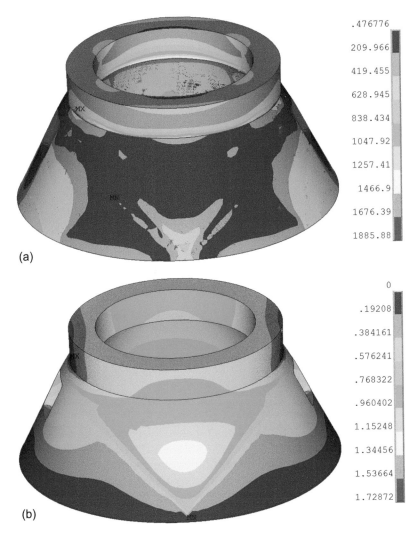

FIGURE 19.7
Final stresses and deformation of the model with damage. (a) vM stresses with damage (MPa). (b) Deformation with damage.

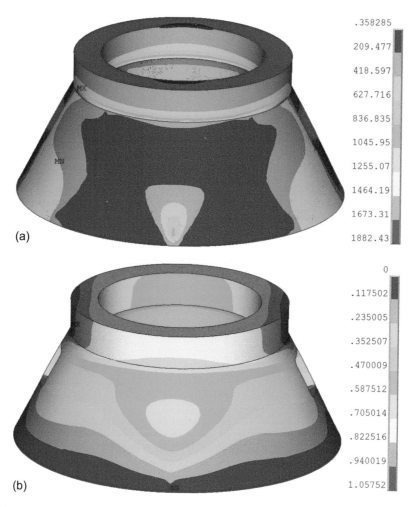

FIGURE 19.8
Final stresses and deformation of the model without damage. (a) vM stresses with damage (MPa). (b) Deformation with damage.

that in the model without damage, and their differences increase with the increase of the loading.

Figure 19.10 illustrates the damage of the composites in the outer and inner surfaces. The damage is initiated at the bottom and the corners of the open doors and grows in two symmetrical branches. The damage starts slowly at the inner surface but grows faster than the outer surface. At the final stage, complete damage (marked as 2) occurs at the corners of the open doors in the inner surface.

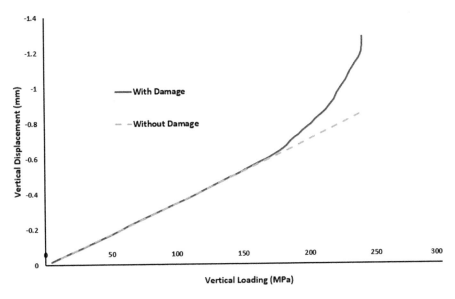

FIGURE 19.9
Displacement history of one node located on the top surface.

19.3.5 Discussion

The damage process of the composite structure was simulated in ANSYS. Compared to the model without damage, the model with damage has much large deformation under the same loading, and the differences between the two models increase with the increase of the loading, which is close to the reality.

For the solution part, the model with damage needs many iterations to reach the convergence for each substep. On the other hand, the model without damage finishes the whole computation in a few steps.

The composite faces with 0° are in the axial direction. Thus, the local coordinate system with X in the vertical direction was defined as the element coordinate system. Therefore, the element X axis is determined from the projection of the local coordinate system on the shell element.

19.3.6 Summary

The composite in the composite structure failure test is modeled as a sandwich; and its damage was studied in ANSYS190. With the force loading, its deformation in the damage stage increases much faster than that of the model without damage.

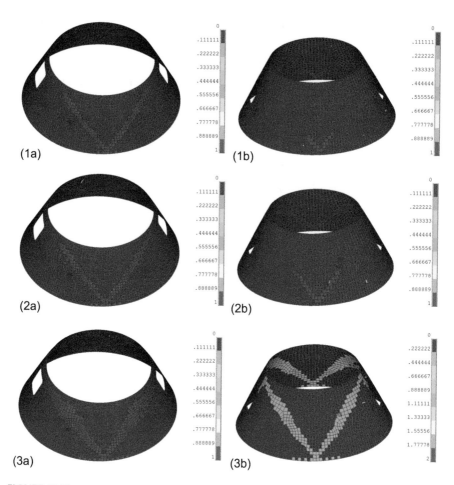

FIGURE 19.10
Damage of the composites in the outer and inner surfaces under different loading scales. (1a) Outer side under loading 85%, (1b) inner side under loading 85%. (2a) Outer side under loading 90%, (2b) inner side under loading 90%. (3a) Outer side under loading 100%, (3b) inner side under loading 100%.

19.4 Simulation of Crack Growth in Single Leg Bending Problem

Composite Structure has two issues that need to be addressed: (1) the damage, which was discussed in Section 19.3 and (2) delamination. Here, the single leg bending (SLB) problem was studied using VCCT, which sets an example for the study of composite delamination.

19.4.1 Finite Element Model

One SLB problem was built in ANSYS [4] (see Figure 19.11). The whole model was meshed with PLANE182 (plane strain), while the interface was modeled with the Cohesive Zone element (CZM)202 with plane strain. The initial crack was defined between the top area and bottom area.

19.4.2 Material Properties

The material of the composite was defined as anisotropic by

```
MP,EX,4,135.3E3      !* Material properties of lamina
MP,EY,4,9.0E3
MP,EZ,4,9.0E3
MP,GXY,4,4.5E3
MP,GYZ,4,4.5E3
MP,GXZ,4,3.3E3
MP,PRXY,4,0.24
MP,PRXZ,4,0.24
MP,PRYZ,4,0.46
```

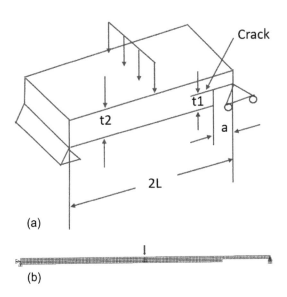

FIGURE 19.11
SLB problem and its finite element model (t1 = 2.032 mm, t2 = 4.064 mm, a = 34.3 mm, 2L = 177.8 mm). (a) SLB problem. (b) Finite element model.

Linear fracture criterion was specified for the crack growth with an energy release rate 0.05 for critical Mode I, Mode II, and Mode III as

```
tb,cgcr,1,,3,Linear      !* Linear fracture criterion
tbdata,1,0.05,0.05,0.05
```

19.4.3 Crack Definition

The crack was defined in the model using CINT command as

```
cint,new,1
cint,type,vcct
cint,ctnc,crack1
cint,norm,12,2
cint,ncon,6        ! number of contours
cint,symm,off      ! symmetry off
```

The crack growth was specified as

```
cgrow,new,1
cgrow,method,vcct
cgrow,cid,1        ! cint ID VCCT calculation
cgrow,cpath,cpath1 ! crack path
```

19.4.4 Boundary Conditions and Loadings

The center of the left end was fixed in the vertical direction, and the right end was constrained with all degrees of freedom. The top center was loaded with displacement 5 mm (see Figure 19.11).

19.4.5 Results

The final deformation of the model was illustrated in Figure 19.12. It clearly shows that crack grows while the model is under bending. The length of crack growth 82.9 mm was obtained using command prci,1,,CEXT in the post process. Command PRCINT,1,,VCCT prints out the energy release rate of the last step,

$$G_1 = 0.0274 \text{ MPa mm}$$

$$G_2 = 0.0233 \text{ MPa mm}$$

FIGURE 19.12
Deformation of the model (Dmax = 10 mm).

$$G_3 = 0.0 \text{ MPa mm}$$

$$G_{tot} = 0.0507 \text{ MPa mm}$$

Obviously, the energy release rate of Mode I and Mode II are very close, while it is zero for Mode III.

Figure 19.13 plots the reaction force with the displacement loading. At the starting stage, the reaction force increases with the displacement loading. After the crack growth starts, the reaction force drops significantly and remains while the crack growth continues. When the crack growth stops, the reaction force increases linearly with an increase of the displacement loading.

19.4.6 Discussion

The crack growth of the SLB problem was simulated using VCCT in ANSYS190. Unlike the double cantilever beam (DCB) problem in which Mode I dominates, both Mode I and Mode II play a close role in the crack growth of SLB.

ANSYS supports the cohesive-zone model (CZM) method and the virtual crack closure technique (VCCT) method. The SLB problem can be studied using the CZM method as well.

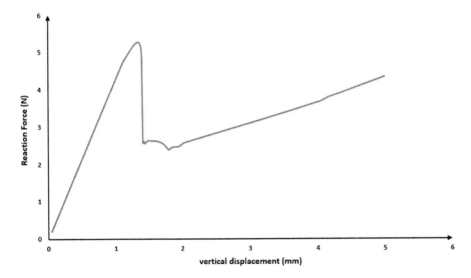

FIGURE 19.13
Reaction force with the displacement loading.

19.4.7 Summary

The SLB was modeled in ANSYS190, and its crack growth was simulated using VCCT. The results indicate that both Mode I and Mode II are involved in the crack growth.

References

1. Jones, R.M., *Mechanics of Composite Materials*, CRC Press, 1998.
2. ANSYS190 Help Documentation in the help page of product ANSYS190.
3. Frappier, A.M., An investigation of composite failure analyses and damage evolution in finite element models, Purdue University, Master Theses, 2006.
4. Krueger, R., Shivakumar, K., Raju, I.S., Fracture mechanics analyses for interface crack problems – a review, 54th AIAA/ASME/ASCE/AHS/ASC Structures, Structural Dynamics, and Materials Conference, Structures, Structural Dynamics, and Materials and Co-located Conferences, Boston, Massachusetts, 2013.

20

Functionally Graded Materials

Functionally graded materials (FGMs), such as bones and teeth, exist in the human body; these materials are designed to meet functional requirements. Functionally graded materials are imitated from nature to solve engineering problems. Different kinds of fabrication processes are developed to produce functionally graded materials. Chapter 20 discusses functionally graded materials, including their introduction in Section 20.1, their modeling using ANSYS in Section 20.2, and one example of an FGM in Section 20.3.

20.1 Introduction of Functionally Graded Materials

As revolutionary materials, functionally graded materials are a set of advanced materials with properties changing over a varying dimension. The difference between FGMs and composite materials is that sharp interfaces exist in composites, which cause fracture and failure. On the other hand, FGMs have a smooth transition from one material to another by using a gradient interface.

Generally, FGMs are classified into two groups: (1) thin FGMs with a relatively thin surface coating and (2) bulk FGMs with volumes of materials requiring more intensive processes [1]. Thin section or surface coating FGMs are fabricated using physical or chemical vapor deposition (PVD/CVD), plasma spraying, or self-propagating high-temperature synthesis (SHS), while bulk FGMs are produced using powder metallurgy, centrifugal casting method, and solid freeform technology.

With improvements in fabrication processes, the overall process cost for manufacturing FGMs has dramatically reduced. Thus, the application of FGMs has expanded in many fields such as aerospace, automobile industry, medicine, sports, energy, sensors, and optoelectronics.

20.2 Material Model of Functionally Graded Materials

The material properties of functionally graded materials are a function of location, which can be modeled in ANSYS through two different approaches. The first approach, the interpolation algorithm, uses TBIN, ALGO to define the material properties as

```
tb,elas,1            ! defines the material properties
*do, i, np           ! np is the number of supporting points
! assign Young's modulus Enp for the supporting points
  E%i%= Enp(i)
!(xxi, yyi) is the location of supporting point i
  Tbfield, xcor,    xx%i%
  tbfield, ycor,    yy%i%
  tbdata,   1,    E%i% , nu ! nu is Poisson's ratio
*enddo
tbin,algo
```

In the second approach, every element has its own material identity with material parameters computed from the location of the element center. The approach is implemented using DO LOOP:

```
! define the material properties
*do, i, 1, ne    ! ne is the number of elements
  E%i%=Ep        ! Ep is Young's modulus of element i
  MP,EX,i,E%i%
  MP,NUXY,i,nu   ! nu is Poisson's ratio of element i
*enddo
```

The difference between these two approaches is that the first approach has only one material identity, but the second approach has as many material identities as the number of elements. Therefore, the first approach has an advantage: it is insensitive to meshing, while the second approach is sensitive to meshing. With a finer meshing and more elements, the second approach gets closer to the first approach.

20.3 Simulation of a Spur Gear Fabricated Using Functionally Graded Materials

Spur gears have a wide range of applications in power transmission. The hardworking conditions of the spur gears require them to resist tooth deflection and stress concentration, which demand that the gear material has sufficient strength. A functionally graded material is a revolutionary material and was selected for fabricating a spur gear in this study; the gear was simulated using two different methods. The results can be used as a guideline for the mechanical design of spur gears.

20.3.1 Finite Element Model

A spur gear was meshed with PLANE182, assuming plane stress (Figure 20.1) [2].

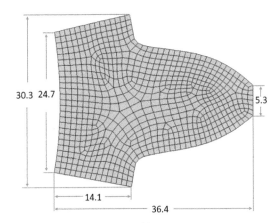

FIGURE 20.1
Finite element model of a spur gear (all dimensions in mm).

20.3.2 Material Properties

The spur gear was assumed to be made of a functionally graded material, following the exponential law [2, 3]:

$$E = 2.01 \times 10^5 e^{0.0097(x-14.9)} \qquad (20.1)$$

The material properties of the spur gear were defined in two different ways. The first one was the interpolation algorithm using command TBIN, ALGO. The material of the gear was defined as

```
tb,elas,1        ! define the material properties
*do, i, 1, 11
! assign Young's modulus for the supporting points
E%i%=2.01e5*exp(0.0097*(i-1)*2)
tbfield, xcor,   (i-1)*2+14.9
tbdata,  1,      E%i% , 0.0
*enddo
```

The second approach defined a number of material models, which have individual Young's moduli calculated by equation (20.1) using different x values. These material models were assigned to the appropriate area of the spur gear according to the corresponding x value.

```
! define the material properties
*do, i, 1, 6
! assign Young's modulus for the supporting points
E%i%=2.01e5*exp(0.0097*(i-1)*4)
MP,EX,i,E%i%
MP,NUXY,i,0.
*enddo
```

Both approaches have a very close distribution of Young's modulus (see Figure 20.2).

20.3.3 Loadings and Boundary Conditions

The inner rim of the gear tooth and the two radial lines were constrained in both the x and y directions. A tangential force 113.68 N was applied at the tip of the gear tooth [2]. To reduce stress concentration, the force was loaded equally on two nodes (see Figure 20.3).

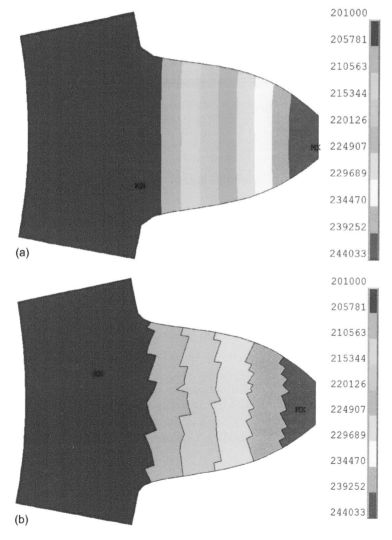

FIGURE 20.2
Young's modulus distribution of the spur gear through: (a) TBField and (b) equivalent materials.

Functionally Graded Materials

20.3.4 Results

Figure 20.4 depicts the deformation of the gear. Both approaches result in almost the same deflection, about 8.5×10^{-5} mm. The vM stresses of the gear are plotted in Figure 20.5. The maximum stress occurs at the loading area due to stress concentration, about 0.92 MPa. The stress results of the two approaches are very close.

20.3.5 Discussion

The spur gear made using a functionally graded material was simulated using two different methods. Both methods resulted in very close displacements and stresses, which validate the interpolation algorithm of ANSYS.

The gear defined by TBIN, ALGO has continuous Young's modulus with one material identity. On the contrary, Young's modulus obtained by the equivalent material model is discontinuous and defined by a number of material identities.

20.3.6 Summary

The spur gear was modeled as a functionally graded material by two different methods – the interpolation algorithm and the equivalent material

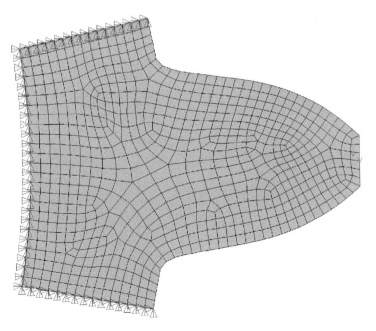

FIGURE 20.3
Boundary conditions and loadings.

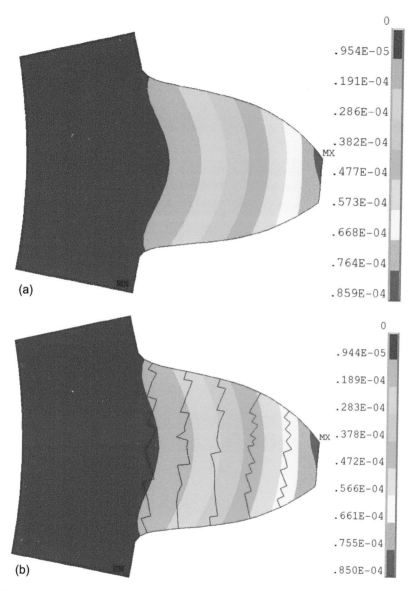

FIGURE 20.4
Deformation of the spur gear (mm): (a) TBField and (b) equivalent materials.

model. Both showed very close results in terms of deformation and stress distribution. Unlike the equivalent material model, the interpolation algorithm defined the functionally graded material with one material identity and showed a continuous distribution of Young's modulus.

Functionally Graded Materials

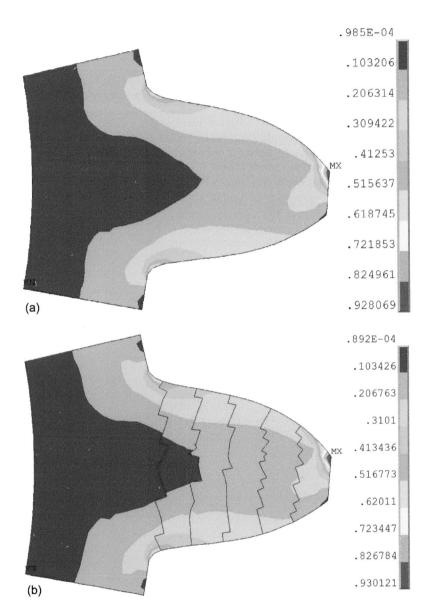

FIGURE 20.5
vM stresses of the spur gear (MPa): (a) TBField and (b) equivalent materials.

References

1. Mahamood, R.M., Akinlabi, E.T., Shukla, M., Pityana, S., Functionally graded material: an overview, Proceedings of the World Congress on Engineering 2012, Vol. III, WCE2012, July 2012, London, UK, 2012.
2. Kumar, R., Sarda, A.A., Analysis of functionally graded material spur gear under static loading condition, *Advance Physics Letter*, Vol. 3, 2016, pp. 1–6.
3. Pawar, P.B., Utpat, A.A., Analysis of composite material spur gear under static loading condition, 4th International Conference on Materials Processing and Characterization, Vol. 2, 2015, pp. 2968–2974.

21

Shape Memory Alloys

Shape memory alloys (SMAs) refer to a special type of alloy that can recover its initial shape after severe deformations. Their unique features make them widely applicable in industry – for orthodontic wires, eyeglasses, and stents in the biomedical field; for actuators and connectors in the mechanical field; and for shape control and vibration in the structural field. In Chapter 21, the structure of SMAs and the corresponding material models are introduced, followed by SMAs' application for an orthodontic wire and a vacuum-tight shape memory flange.

21.1 Structure of SMAs and Various Material Models

21.1.1 Structure of SMAs

SMAs have two different phases: (1) the high-temperature phase named austenite (A) and (2) the low-temperature one called martensite (M). Both have different crystal structures and therefore different properties [1]. Figure 21.1 illustrates the crystal structure of austenite that is generally cubic. On the other hand, martensite exists in two forms: twinned martensite (see Figure 21.2a) and detwinned martensite (see Figure 21.2b). The phase transformation from austenite to martensite and vice versa generates the unique features of SMAs, which are as follows.

21.1.1.1 Superelasticity

The superelasticity of SMAs relates to stress-induced transformation at a temperature above the austenitic finish temperature (A_f. Consider a loading path (A→B→C→D→E→F→A) as shown in Figure 21.3. When a mechanical load is applied up to B, the SMA has an elastic deformation in the austenite phase. After the load reaches B (the detwinning start stress, σ^M) (see Figure 21.4), a phase transformation starts accompanying the transformation strain. The transformation continues until C (σ^{Ms}), which is the end of the transformation. From C to D, the SMA has an elastic deformation in the detwinned martensite phase. In the unloading stage of D→E, the SMA remains elastic. At point E, the stress reaches σ^{As} that causes the martensite to return to

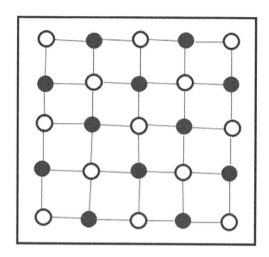

FIGURE 21.1
Crystal structure of austenite.

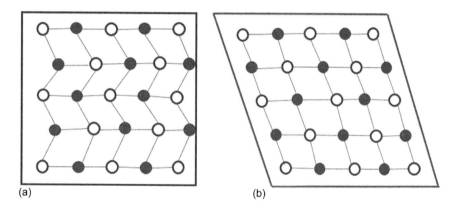

FIGURE 21.2
Structure of martensite: (a) twinned martensite and (b) detwinned martensite.

austenite. The whole phase transformation occurs between E and F. F is the end of the phase transformation with stress σ^{Af}. At the final unloading stage (F→A), the SMA returns to its origin elastically.

Obviously, the superelasticity of SMAs is much different from the large deformation of elastic materials. For elastic materials, the unloading path always retraces the loading path. However, the SMAs' unloading path differs from their loading path because σ^{Mf} is different from σ^{As}, and σ^{Ms} is not equal to σ^{Af}, which causes the phase transformation to occur at different stress levels.

Shape Memory Alloys

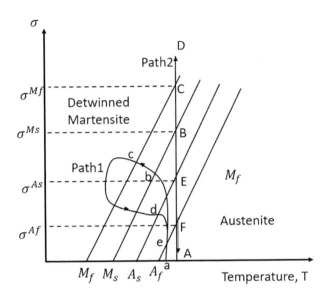

FIGURE 21.3
Stresses of SMAs with temperature [1].

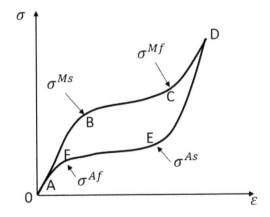

FIGURE 21.4
Various stresses of SMAs with strains.

21.1.1.2 Shape Memory Effect

Assume a loading path (A→B→C→D→E→A) as shown in Figure 21.5. A→B is a cooling stage. When the temperature drops below the forward transformation temperature (M_s and M_f), the phase of the SMA changes from the parent phase to twinned martensite. In the loading stage from B to C, when the stress state of the SMA is above the starting stress level σ_s, reorientation starts. The detwinning process continues with an increase in loading and

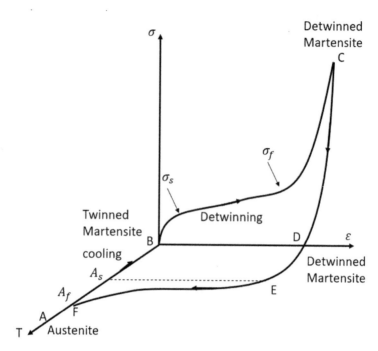

FIGURE 21.5
Shape memory effect [1].

ends at a stress level σ_f. Detwinned martensite remains and deforms elastically in the unloading stage from C to E. The heating process is D→E→A. When the temperature reaches A_s (point E), the transformation from detwinned martensite to austenite starts and then completes at A_f (point A). Thus, the original shape of the SMA is regained.

21.1.2 Various SMA Material Models

The aforementioned SMA features can be modeled in ANSYS by using two different material models [2]: SMA model for superelasticity and SMA material model with the shape memory effect.

21.1.2.1 SMA Model for Superelasticity

The elastic strain ε_e is obtained by subtracting the transformation strain ε_{tr} from the total strain ε:

$$\varepsilon_e = \varepsilon - \varepsilon_{tr} \qquad (21.1)$$

Thus, the stress σ is calculated by

$$\sigma = D : \varepsilon_e = D : (\varepsilon - \varepsilon_{tr}) \qquad (21.2)$$

Shape Memory Alloys

The Drucker–Prager model is used to describe the phase transformation:

$$F = q + 3\alpha p \tag{21.3}$$

where

$$q = \sqrt{\frac{3}{2} S : S} \tag{21.4}$$

$$S = \sigma - p\mathbf{1} \tag{21.5}$$

$$p = \frac{1}{3} \sigma : \mathbf{1} \tag{21.6}$$

α is the material parameter that takes into account the different material behaviors of tension and compression.

Thus, the evolution of the martensite fraction, ξ_s, is a function of F

$$\dot{\xi}_s = \begin{cases} -H^{AS}(1-\xi_s) \dfrac{\dot{F}}{F - R_f^{AS}} & A \to S \\ H^{SA} \xi_s \dfrac{\dot{F}}{F - R_f^{SA}} & S \to A \end{cases} \tag{21.7}$$

where

$$R_f^{AS} = \sigma_f^{AS}(1+\alpha) \tag{21.8}$$

$$R_f^{SA} = \sigma_f^{SA}(1+\alpha) \tag{21.9}$$

$$H^{AS} = \begin{cases} 1 & \text{if } R_s^{AS} < F < R_f^{AS} \text{ and } \dot{F} > 0 \\ 0 & \text{otherwise} \end{cases} \tag{21.10}$$

$$H^{SA} = \begin{cases} 1 & \text{if } R_f^{SA} < F < R_s^{SA} \text{ and } \dot{F} < 0 \\ 0 & \text{otherwise} \end{cases} \tag{21.11}$$

$$R_s^{AS} = \sigma_s^{AS}(1+\alpha) \tag{21.12}$$

$$R_s^{SA} = \sigma_s^{SA}(1+\alpha) \tag{21.13}$$

Therefore, the transformation strain tensor, ε_{tr}, is expressed in terms of ξ_s and yield function F:

$$\dot{\varepsilon}_{tr} = \dot{\xi}_s \bar{\varepsilon}_L \frac{\partial F}{\partial \sigma} \tag{21.14}$$

$\bar{\varepsilon}_L$ is the maximum residual strain.

Equations (21.1)–(21.14) include these material parameters: σ_s^{AS}, σ_f^{AS}, σ_s^{SA}, σ_f^{SA}, $\bar{\varepsilon}_L$, and α. Therefore, the SMA model for superelasticity is defined in ANSYS by commands:

```
TB, SMA, , , ,SUPE
TBDATA, 1, σ_s^AS, σ_f^AS, σ_s^SA, σ_f^SA, ε̄_L, α
```

21.1.2.2 SMA Model with Shape Memory Effect

Similar to the SMA with superelasticity, the total stress in the SMA with the shape memory effect is calculated by

$$\sigma = D:(\varepsilon - \varepsilon_{tr}) \qquad (21.15)$$

where D is a linear function of ε_{tr}

$$D = \frac{\varepsilon_{tr}}{\epsilon_L}(D_M - D_A) + D_A \qquad (21.16)$$

When the material is in its austenite phase, $D = D_A$; when the material is in its martensite phase, $D = D_M$.

A Prager-Lode type function is selected for the yield function to take into account the asymmetric behavior of the SMA under tension and compression:

$$F(X_{tr}) = \sqrt{2J_2} + m\frac{J_3}{J_2} - R \qquad (21.17)$$

where

$$J_2 = \frac{1}{2}\left(X_{tr}^2 : 1\right) \qquad (21.18)$$

$$J_3 = \frac{1}{3}\left(X_{tr}^3 : 1\right) \qquad (21.19)$$

$$X_{tr} = \sigma' - \left[\tau_M(T) + h\|\varepsilon_{tr}\| + \gamma\right]\frac{\varepsilon_{tr}}{\|\varepsilon_{tr}\|} \qquad (21.20)$$

$$\tau_M(T) = \langle \beta(T - T_0) \rangle^+ \qquad (21.21)$$

β is the temperature scaling parameter;
T is the working temperature;
T_0 is the temperature M_f below which no twinned martensite occurs;
h is the hardening parameter; and
γ is given by

$$\begin{cases} \gamma = 0 & \text{if } 0 < \|\varepsilon_{tr}\| < \varepsilon_L \\ \gamma \geq 0 & \text{if } \|\varepsilon_{tr}\| = \varepsilon_L \end{cases} \qquad (21.22)$$

ε_L is the maximum transformation strain;
m is Lode dependency parameter; and
R is the elastic limit.

The evolution of ε_{tr} is related with the yield function F,

$$\varepsilon_{tr} = \varepsilon_{tr}(n) + \Delta \zeta \frac{\partial F}{\partial \sigma} \tag{21.23}$$

Equations (21.15)–(21.23) cover seven material parameters: h, T_0, R, β, ε_L, Em, and m. Thus, the SMA model with the shape memory effect is defined in ANSYS by commands:

```
TB,SMA,,,,MEFF
TBDATA, 1, h, T0, R, β, ϵ_L
TBDATA, 6, Em, m
```

21.1.3 Definition of Material Parameters

21.1.3.1 SMAs with Superelasticity

Some material parameters of SMAs with superelasticity, such as E, σ_s^{AS}, σ_f^{AS}, σ_s^{SA}, σ_f^{SA}, and ε_L, can be determined by the strain–stress curve of the uniaxial tension test illustrated in Figure 21.6. The material parameter α specifies the material response under tension and compression. If tension and compression have the same response, then $\alpha=0$. Generally, α is defined by the starting stress of the austenite to martensite phase transformation under tension and compression as

$$\alpha = \frac{\sigma_c^{AS} - \sigma_t^{AS}}{\sigma_c^{AS} + \sigma_t^{AS}} \tag{21.24}$$

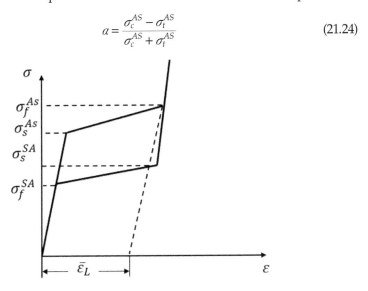

FIGURE 21.6
Strain–stress of SMAs with superelasticity.

where

σ_c^{AS} is the initial stress of the austenite to martensite phase transformation under compression.

σ_t^{AS} is the initial stress of the austenite to martensite phase transformation under tension.

21.1.3.2 SMAs with Shape Memory Effect

E_A and E_m are determined by the slope of the elastic stages in the SMA strain–stress curve (see Figure 21.7). H is the slope of the transformation strain–stress curve, and ε_L is the maximum transformation strain (see Figure 21.8). T_0 is M_f and β is the slope of the temperature–stress plot of the SMA (see Figure 21.9).

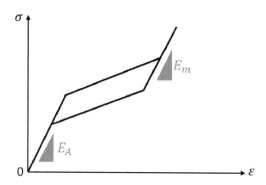

FIGURE 21.7
Strain–stress curve of SMAs.

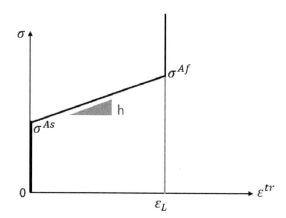

FIGURE 21.8
Transformation strain–stress curve of SMAs.

Shape Memory Alloys

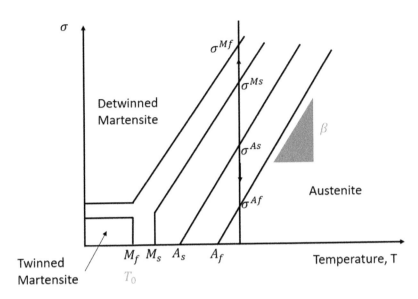

FIGURE 21.9
Temperature–stress curve of SMAs [1].

R and m are determined by tension limit σ_t and compression limit σ_c by

$$R = 2\sqrt{2/3}\,\frac{\sigma_c \sigma_t}{\sigma_c + \sigma_t} \quad (21.25)$$

$$m = \sqrt{27/2}\,\frac{\sigma_c - \sigma_t}{\sigma_c + \sigma_t} \quad (21.26)$$

21.2 Simulation of Orthodontic Wire

Biocompatibility studies indicate that SMA NiTi alloys have low cytotoxicity and genotoxicity [3, 4], which allows for the possibility of their application in orthodontics. The orthodontic wire (see Figure 21.10), as a part of a fixed orthodontic appliance, delivers a force on teeth that permits tooth movement without tissue damage. In the early stage of orthodontic treatment, wires made using SMA NiTi alloys are used for teeth leveling and aligning. In this study, a wire made using SMA NiTi alloys was simulated using the ANSYS SMA model for superelasticity.

21.2.1 Finite Element Model

Because of its symmetry, half of the finite element model was developed in ANSYS190 (see Figure 21.11) [5]. The model was meshed using SOLID185 with 1,856 elements and 2,925 nodes.

FIGURE 21.10
Orthodontic wires (Aleksandra Gigowska ©*123RF.com*).

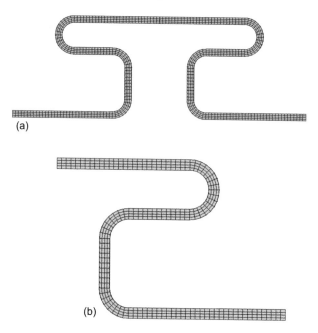

FIGURE 21.11
Finite element model of a wire: (a) full model and (b) half model with symmetrical condition.

21.2.2 Material Properties

The wire was modeled using the SMA material model with superelasticity. The material parameters are listed in Table 21.1 [5].

TABLE 21.1
Material parameters of the wire

E_A (MPa)	v	σ_s^{AS} (MPa)	σ_f^{AS} (MPa)	σ_s^{SA} (MPa)	σ_f^{SA} (MPa)	$\bar{\varepsilon}_L$	α
50,000	0.36	500	500	300	300	0.07	0.136

21.2.3 Loadings and Boundary Conditions

The whole simulation was completed with two loading steps. In the first step, the right end was pulled along the x direction by 7 mm. Then, it was brought back to the initial position in the second step.

The symmetric surface was constrained with all degrees of freedom as boundary conditions (see Figure 21.12).

21.2.4 Results

Figures 21.13 through 21.15 plot the deformation, von Mises stresses, and the transformation strains of the first step, respectively. The deformation and the von Mises stresses of the second step are presented in Figures 21.16 and 21.17, respectively.

In the first step, the wire has a maximum deformation of 8.1 mm. The maximum vM stress of 927.9 MPa and the maximum transformation strain of 0.061 occur at the same location. In the second step, after the wire returns to its initial position, the deformation reduces to near zero, and the maximum vM stress becomes as low as 3.54 MPa, which validates the superelasticity of the SMA.

Figure 21.18 illustrates the vM stresses and plastic strains with time of element A located in the maximum stress area. At the beginning of the first step,

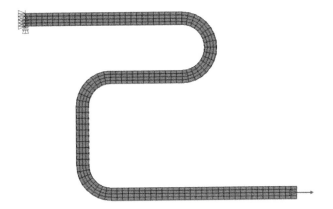

FIGURE 21.12
Boundary condition of orthodontic wire.

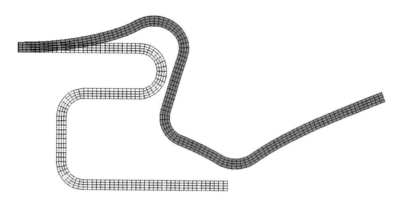

FIGURE 21.13
Deformation of the orthodontic wire at the end of loading stage (DMX=8.1 mm).

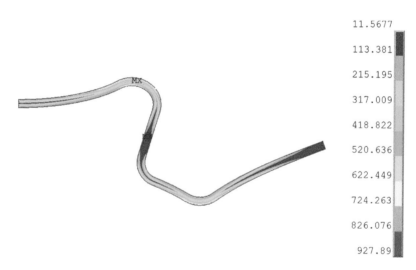

FIGURE 21.14
vM stresses of the orthodontic wire at the end of the loading stage (MPa).

the stress increases linearly without phase transformation. After it reaches phase transformation, the transformation strain increases quickly, while the stress increases flatly. In the early stage of the second step, the material is in full martensite. The stress drops sharply, while the transformation strain remains constant. Then, the material reaches phase transformation. As in the loading stage, the transformation strain drops quickly, while the stress reduces slowly. In the final stage of the second stage, the material is full austenite with zero transformation strain.

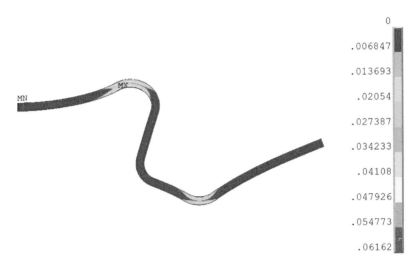

FIGURE 21.15
Transformation strains of the orthodontic wire at the end of the loading stage.

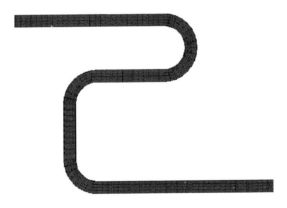

FIGURE 21.16
Deformation at the end of unloading (DMX = 0.2×10^{-3} mm).

21.2.5 Discussion

A three-dimensional finite element model of a wire was simulated using the SMA material model for superelasticity. After the wire returns to its starting position, the stress and transformation strain are near zero, which demonstrates the superelasticity of SMAs.

However, the stress and transformation strain go through different paths in the first loading and second unloading steps, which are different from those of general elastic materials.

FIGURE 21.17
vM stresses of the wire at the end of unloading (MPa).

21.2.6 Summary

An orthodontic wire was studied using the SMA material model with superelasticity, which exhibited the superelasticity of SMAs.

21.3 Simulation of a Vacuum-Tight Shape Memory Flange

A vacuum-tight shape memory flange was designed for the JET in-vessel inspection system [6], composed of an SMA sleeve and the base connecting the flange to the other apparatus. A silica rod connected a chamber requiring high vacuum (see Figure 21.19). Initially, the diameter of the sleeve was smaller than that of the rod. Under an external force, the sleeve was enlarged to a greater diameter than that of the rod. Thus, the flange was placed around the rod. After heating the flange, the SMA sleeve underwent transformation from martensite to austenite and regained its shape; this caused close contact between the sleeve and the rod, ensuring high vacuum. The whole process was simulated in ANSYS190.

21.3.1 Finite Element Model

Due to symmetry in two directions, a quarter of the finite element model was built in ANSYS190, including the SMA sleeve, the base, and the rod (see Figure 21.20). The model was meshed using SOLID185 with 2,291 elements and 3,403 nodes.

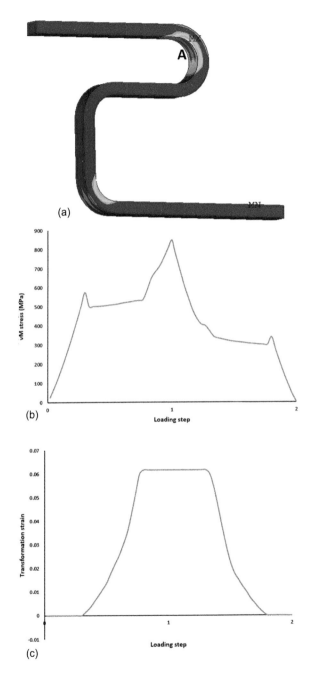

FIGURE 21.18
Time history of stress and transformation strain in element A: (a) element A, (b) stress in element A versus time, and (c) transformation strain versus time.

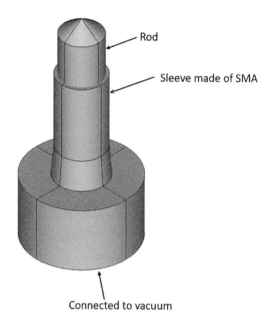

FIGURE 21.19
Illustration of a vacuum-tight shape memory flange.

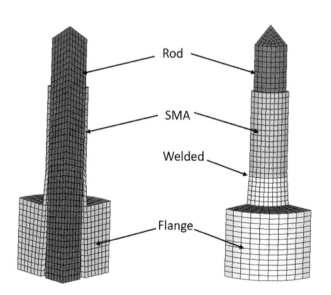

FIGURE 21.20
Finite element model of an SMA flange.

TABLE 21.2
Material parameters of the sleeve

E_A (MPa)	v	H	T_0	R	β	$\bar{e}_L$	E_m (MPa)	m
60,000	0.36	1000	223	50	2.1	0.04	45,000	0.0

21.3.2 Material Properties

The sleeve was modeled with the SMA material model with the shape memory effect, because it involves temperature change. The material parameters are listed in Table 21.2 [6].

The rod and the base were assumed to be linear elastic with Young's moduli of 74 MPa and 53 MPa, respectively [6].

21.3.3 Contact

The sleeve was connected to the base through welding. Thus, they were glued together, sharing common nodes. The contact between the flange and the rod was defined as standard with TARGE170 and CONTA174.

21.3.4 Loadings and Boundary Conditions

The loading went through the following three steps:

(1) A pressure of 22 MPa was applied on the sleeve's internal face to enlarge the SMA sleeve while deactivating the contact between the rod and the sleeve. In this stage, the temperature was maintained at 223 K;
(2) The rod–sleeve contact was activated and the applied pressure was removed. The temperature remained at 223 K in this stage; and
(3) The temperature was increased to 300 K to induce phase transformation.

The bottom surface was constrained with all degrees of freedom as boundary conditions.

21.3.5 Solutions

The SMA material model with the shape memory effect was employed. Thus, the full Newton–Raphson option with unsymmetrical metrices of elements was selected in the solution setting. In addition, contact existed in the model. To reach convergence, a very small time-step of 0.001 was selected for the second and third steps.

21.3.6 Results

Figures 21.21 through 21.23 plot the displacements and von Mises stresses for each step. Transformation strains of the second and third steps are presented in Figures 21.24 and 21.25, respectively. At the end of the first step, the sleeve enlarges with a maximum displacement of 5.17 mm. The maximum stress of 362.8 MPa occurs at the connection of the sleeve and the base. In the second step, after the pressure loaded in the first step is removed, the sleeve still has a deformation as high as 4.78 mm, and the maximum stress of 260.7 MPa still exists at the bonding between the sleeve and the base. The remaining deformation of the sleeve due to the existing transformation strain (see Figure 21.24) causes the deformation of the base because of the continuous displacement at the connection between the sleeve and the base. That is why a big stress occurs at the interface between the sleeve and base. In the third stage, heating induces phase transformation. Thus, the deformation drops significantly to 0.85 mm, and the maximum transformation strain reduces from 0.032 in the second step to 0.0058 in the third step. Therefore, in this step, the stresses in the sleeve are primarily caused by the contact between the rod and the sleeve with a maximum stress of 145.4 MPa. The contact pressure is illustrated in Figure 21.26.

Figure 21.27 describes the vM stress with time in element B of the base, which indicates that its stress increases to 360 MPa in the first step and gradually drops to 250 MPa in the second step. At the beginning of the third step, it quickly decreases to around 20 MPa. It does not change much for the remainder of the third step. The variation of the phase transformation strain with stress in element A of the sleeve is plotted in Figure 21.28. It clearly shows that the stress increases in the loading stage in the first step and drops in the unloading stage in the second step. In the third step, with phase transformation, the transformation strain decreases gradually with near-zero stress. When the sleeve starts to contact the rod, its stress quickly increases.

21.3.7 Discussion

A three-dimensional finite element model of a vacuum-tight flange was built. A part of the flange was modeled using the SMA material model with the shape memory effect. The phase transformation due to heating was simulated, which demonstrated the shape memory effect of SMAs.

This study did not take thermal strain into account because in the model, thermal strain is much less than transformation strain. In addition, phase transformation is strongly nonlinear and its governing equations are unsymmetrical. Solution convergence requires NROPT, UNSYM, and a very small time-step defined in the solution setting.

In this study, the contact between the sleeve and the rod was deactivated and activated using the element birth and death technology, which is widely used in applications such as the modeling of soil excavation.

Shape Memory Alloys

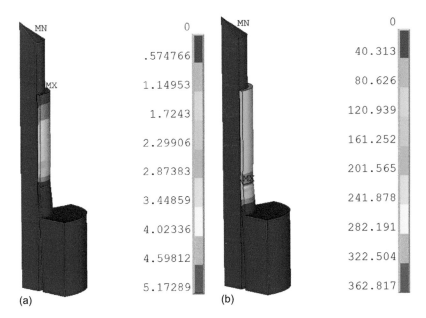

FIGURE 21.21
Results at the end of the first step: (a) displacement (mm) and (b) vM stress (MPa).

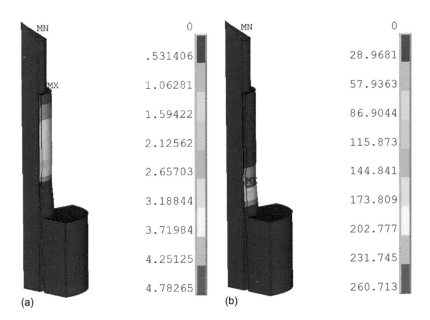

FIGURE 21.22
Results at the end of the second step: (a) displacement (mm) and (b) vM stress (MPa).

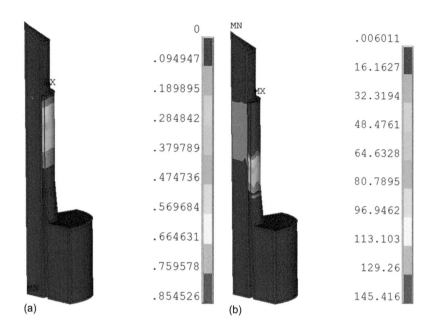

FIGURE 21.23
Results at the end of the third step: (a) displacement (mm) and (b) vM stress (MPa).

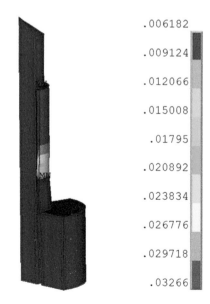

FIGURE 21.24
Transformation strains at the end of step 2.

Shape Memory Alloys

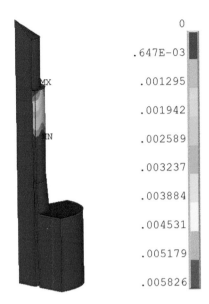

FIGURE 21.25
Transformation strains at the end of step 3.

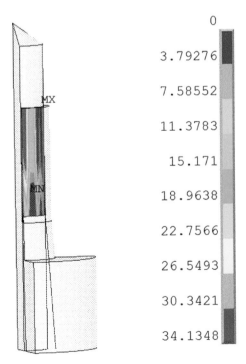

FIGURE 21.26
Contact pressure in the last step (MPa).

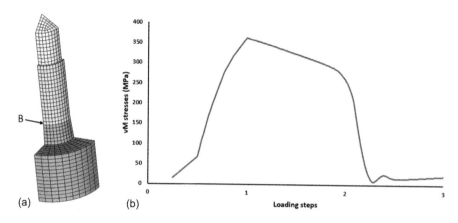

FIGURE 21.27
vM stress in element B with time: (a) element B (b) stress of element B at the different loading steps.

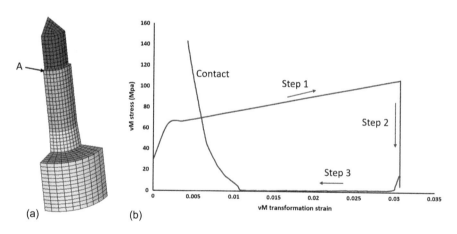

FIGURE 21.28
Phase transformation strain versus stress in element A: (a) element A (b) variation of stress of element A with transformation strain.

21.3.8 Summary

A vacuum-tight flange was simulated using the SMA material model with the shape memory effect, which demonstrates the shape memory effect of SMAs.

References

1. Lagoudas, D.C., *Shape Memory Alloys*, Springer, 2008.
2. ANSYS190 Help Documentation in the help page of product ANSYS190.

3. Es-Souni, M., Es-Souni, M., Fischer-Brandies, H., Assessing the biocompatibility of NiTi shape memory alloys used for medical applications, *Analytical and Bioanalytical Chemistry*, Vol. 381, 2005, pp. 557–567.
4. Rahilly, G., Price, N., Nickel allergy and orthodontics, *Journal of Orthodontics*, Vol. 30, 2003, pp. 171–174.
5. Auricchio, F., A robust integration-algorithm for a finite-strain shape-memory-alloy superelastic model, *International Journal of Plasticity*, Vol. 17, 2001, pp. 971–990.
6. Auricchio, F., Petrini, L., A three-dimensional model describing stress-temperature induced solid phase transformations: solution algorithm and boundary value problems, *International Journal for Numerical Methods in Engineering*, Vol. 61, 2004, pp. 807–836.

22

Simulation of Piezoelectricity

Chapter 22 aims to model piezoelectricity. After introducing piezoelectricity, including the modeling of piezoelectricity in ANSYS, a piezoelectric accelerometer is simulated in ANSYS190 in Section 22.4.

22.1 Introduction of Piezoelectricity

In Greek, "piezo" means pressure. The term piezoelectricity refers to electricity resulting from pressure. Piezoelectricity was studied first by the brothers Jacques Curie (1856–1941) and Pierre Curie (1859–1906) [1–2]. They found that tension and compression in certain crystalline minerals such as tourmaline, quartz, topaz, cane sugar, and Rochelle salt generated voltages of opposite polarity. Later, the Curie brothers discovered the converse effect that one of the voltage-generating crystals exposed to an electric field deformed. After that, people paid more and more attention to the research and application of piezoelectric materials; one of the milestones was the discovery of the phenomenon of ferroelectricity by Valasek. The mixed oxide compound barium titanate $BaTiO_3$ was the first commercial ferroelectric material, which entered the market in 1945.

As one of the branches of crystal physics, piezoelectricity has the following modern applications:

- Frequency control and signal processing such as mechanical frequency filters, surface acoustic wave devices, and bulk acoustic wave devices;
- Actuators and motors as a piezoelectric material deforms when an electric field is applied;
- Sound sensors such as piezoelectric microphones and piezoelectric pickups for electric guitars; and
- Smart structures using discrete piezoelectric patches to monitor the response of a structure.

22.2 Structures and Mechanical Behaviors of Piezoelectric Materials

Piezoelectric materials are a special type of dielectrics in which the electrical and elastic fields are coupled. When a dielectric material is loaded with an external electrical field, electric dipoles are generated due to the interaction of the electrical field with the dielectric structure. The electric dipole refers to an electro-neutral unit volume. When the center of a positive charge within a given region is not in the same position as the center of a negative charge within the same region, a dipole moment µ is generated (see Figure 22.1),

$$\mu = qr \tag{22.1}$$

The polarization of a material is defined as the total dipole moment for a unit volume

$$P = \frac{\sum \mu_i}{V} \tag{22.2}$$

where V is the overall volume. Please note that μ_i is a vector sum. When no electric field is loaded, the polar molecules in the material are randomly orientated. Then, the dipole moments cancel out without any net polarization.

When a piezoelectric material is loaded with mechanical stress, the shape of the atomic structure of the crystal deforms. Thus, ions in the structure separate to generate a dipole moment. A net polarization requires that the piezoelectric atomic structure cannot be centrosymmetric, because the dipole formed must not be canceled out by other dipoles. When a piezoelectric material is applied with electric loading, electrical dipoles form to generate a dipole moment and result in deformation.

Ferroelectric materials have spontaneous polarization. Electric dipoles exist in their structure even without the electrical field. A newly grown single crystal has many domains. Each domain is a homogenous area and has individual polarization. All dipole moments in one domain follow the direction of that domain. With many domains along individual polarizations, no overall polarization appears. To have uniform polarization, a ferroelectric material must go through an additional process called poling (see

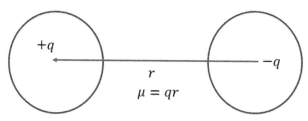

FIGURE 22.1
Dipole moment.

Figure 22.2). As a ferroelectric material passes through its Curie temperature, an electric field is loaded on the material to make its spontaneous polarization aligned in one direction. Therefore, the dipole moments in all the domains point in that direction and have a net dipole moment with the same polarization. After the electric field is removed, most of the dipoles remain nearly aligned. Thus, the material has remnant polarization.

Overall, some conclusions can be made about piezoelectric materials:

(1) The coupling of electric and mechanical fields of piezoelectric materials is caused by the specific asymmetric atomic structure of the lattice and the spontaneous polarization at the microstructure level;

(2) The combination of the ferroelectric properties of piezoelectric materials with the poling process during their manufacture make these materials widely applicable in the modern industry; and

(3) The material properties of piezoelectric materials, including elastic stiffness, dielectric constant, and piezoelectric matrix, are anisotropic due to the polarization of the materials.

22.3 Constitutive Equation of Piezoelectricity

The constitutive equation of piezoelectricity is written in the form of the coupling of elasticity with electrostatics as [4]

$$\begin{Bmatrix} \{T\} \\ \{D\} \end{Bmatrix} = \begin{bmatrix} [C] & [e] \\ [e]^T & -[\varepsilon^S] \end{bmatrix} \begin{Bmatrix} \{S\} \\ -\{E\} \end{Bmatrix} \quad (22.3)$$

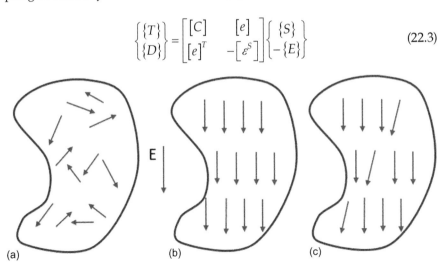

FIGURE 22.2
Structure of ferroelectric materials during poling process [3]: (a) unpoled ceramic, (b) during poling, (c) after poling.

where
- T – stress vector
- C – elasticity matrix
- e – piezoelectric stress matrix
- S – elastic strain vector
- D – electric flux density vector
- ε^S – dielectric matrix
- E – electric field intensity vector

The two rows in equation (22.3) are the constitutive equations for the structural field and electrical field, respectively, except for the coupling terms.

The elasticity matrix [C] has the following form, and is defined in ANSYS using TB, ANEL [4]:

$$[C] = \begin{bmatrix} c_{11} & c_{12} & c_{13} & c_{14} & c_{15} & c_{16} \\ & c_{22} & c_{23} & c_{24} & c_{25} & c_{26} \\ & & c_{33} & c_{34} & c_{35} & c_{36} \\ & & & c_{44} & c_{45} & c_{46} \\ & & & & c_{55} & c_{56} \\ & & & & & c_{66} \end{bmatrix} \tag{22.4}$$

The piezoelectric stress matrix [e] is specified as

$$[e] = \begin{bmatrix} e_{11} & e_{12} & e_{13} \\ e_{21} & e_{22} & e_{23} \\ e_{31} & e_{32} & e_{33} \\ e_{41} & e_{42} & e_{43} \\ e_{51} & e_{52} & e_{53} \\ e_{61} & e_{62} & e_{63} \end{bmatrix} \tag{22.5}$$

and input in ANSYS by using TB, PIEZ [4].

The orthotropic dielectric matrix $[\varepsilon^S]$ is written as

$$[\varepsilon^S] = \begin{bmatrix} \varepsilon_{11} & 0 & 0 \\ 0 & \varepsilon_{22} & 0 \\ 0 & 0 & \varepsilon_{33} \end{bmatrix} \tag{22.6}$$

and described by MP command in ANSYS [4].

Thus, equation (22.3) requires an input of the elasticity matrix [C], the piezoelectric stress matrix [e], and the dielectric matrix $[\varepsilon^S]$ in the finite element models.

22.4 Simulation of Piezoelectric Accelerometer

One of the applications of piezoelectric materials is the piezoelectric accelerometer, which collects the charge generated by a piezoelectric material while it deforms due to seismic mass vibration [5–7]. To achieve better performance of the piezoelectric accelerometer, the finite element method has been used to study it [8–9]. The goal of this study is to find the relation between the lead zirconate titanate (PZT) film thickness and its performance, which provides a base for the mechanical design of the piezoelectric accelerometer.

22.4.1 Finite Element Model

The finite element model of a piezoelectric thin film microaccelerometer was built in ANSYS190 (see Figure 22.3), in which a seismic mass in the center was supported by four suspended symmetric beams [8]. Eight piezoelectric transducers were symmetrically attached on four beams to build the sensing devices. The component dimensions of the accelerometer are listed in Table 22.1. PZT thin films were meshed using SOLID226 with keyoption (1)=1001, while other parts were meshed using SOLID186.

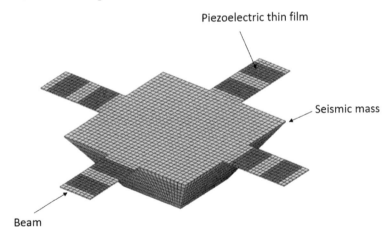

FIGURE 22.3
Finite element model of an accelerometer.

TABLE 22.1
Dimensions of a piezoelectric accelerometer

Length of suspension beam	400 μm
Width of suspension beam	200 μm
Thickness of suspension beam	5 μm
Length of seismic mass	800 μm
Thickness of seismic mass	300 μm
Thickness of PZT film	0.5–3 μm

22.4.2 Material Properties

The seismic mass and the beams were assumed to be linear elastic with a Young's modulus of 190 GPa and a Poisson's ratio of 0.3. PZT was defined as anisotropic with a stiffness coefficient matrix as follows [9]:

$$[C] = \begin{bmatrix} 114.25 & 58.294 & 58.525 & 0 & 0 & 0 \\ 58.294 & 114.25 & 58.525 & 0 & 0 & 0 \\ 58.525 & 58.525 & 98.181 & 0 & 0 & 0 \\ 0 & 0 & 0 & 20.747 & 0 & 0 \\ 0 & 0 & 0 & 0 & 20.747 & 0 \\ 0 & 0 & 0 & 0 & 0 & 26.042 \end{bmatrix} \text{(GPa)} \quad (22.7)$$

The corresponding ANSYS commands are as follows:

```
!Piezoelectric stiffness(Elastic) coefficient
Matrix[c] (µN/µm2) ==> 1e6 N/m² = 1µN/µm²)
c11=11.425e+4
c12=5.8294e+4
c13=5.8525e+4
c22=11.425e+4
c23=5.8525e+4
c33=9.8181e+4
c44=2.0747e+4
c55=2.0747e+4
c66=2.6042e+4
TB,ANEL,2              ! INPUT [C] MATRIX FOR PZT
TBDATA,1,c11,c12,c13
TBDATA,7,c22,c23
TBDATA,12,c33
TBDATA,16,c44
TBDATA,19,c55
TBDATA,21,c66
```

The piezoelectric stress constant matrix was given as [8]

$$[e] = \begin{bmatrix} 0 & 0 & 2.3687 \\ 0 & 0 & 7.6006 \\ 0 & 0 & 16.422 \\ 0 & 0 & 0 \\ 10.249 & 0 & 0 \\ 0 & 0 & 0 \end{bmatrix} \text{(PC/um}^2\text{)} \quad (22.8)$$

Simulation of Piezoelectricity

It was defined in ANSYS as

```
e13=2.3687
e23=7.6006
e33=16.422
e51=10.249
TB,PIEZ,2           ! DEFINE PIEZOELECTRIC Table FOR PZT
TBDATA,3,e13        ! DEFINE MATRIX CONSTANTS
TBDATA,6,e23
TBDATA,9,e33
TBDATA,14,e51
```

The permittivity constant was specified as 10.443×10^{-3} pF/μm, 10.443×10^{-3} pF/μm, and 6.46×10^{-3} pF/μm for X, Y, and Z, respectively [8]. The corresponding ANSYS commands were as follows:

```
! Define dielectric constant [&] (pF/μm) ==> F/m = 1E6 pF/μm)
MP,PERX,2,10.443e-3         ! X DIRECTION PERMITTIVITY
MP,PERY,2,10.443e-3         ! Y DIRECTION PERMITTIVITY
MP,PERZ,2,6.4605e-3         ! Z DIRECTION PERMITTIVITY
```

A local coordinate system 12 was defined for the element coordinates of PZT arranged in the Y direction since PZT is anisotropic.

22.4.3 Boundary Conditions and Loadings

Normally, the accelerometer is attached to the measured body. Thus, the end of each beam was fixed with all degrees of freedom. An acceleration of 1 g was applied on the seismic mass (see Figure 22.4).

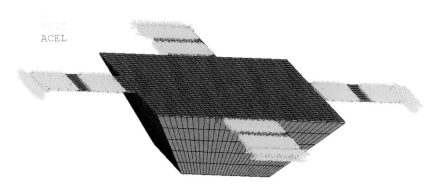

FIGURE 22.4
Boundary conditions and loading.

22.4.4 Results

Figure 22.5 plots the voltage output after the accelerometer was loaded with 1 g acceleration. The four beams have a similar distribution where the voltage increases from -0.735×10^{-3} V at the inside to 0.747×10^{-3} V at the outside. Figure 22.6 illustrates the variation of the PZT film thickness with output

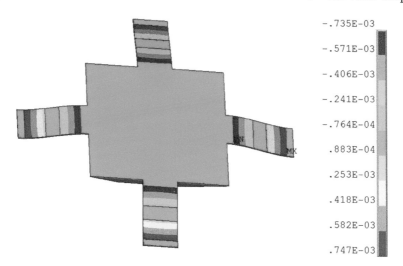

FIGURE 22.5
Voltage output after 1 g acceleration was loaded.

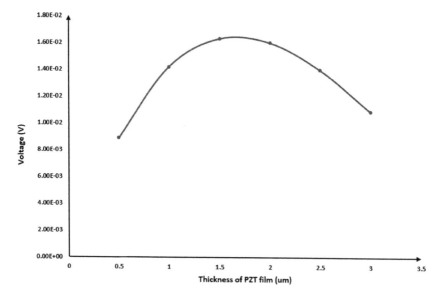

FIGURE 22.6
Variation of voltage with PZT film thickness.

Simulation of Piezoelectricity

voltage; the figure shows that with an increase in the PZT thickness, the voltage output increases to a peak and then drops down.

For comparison, modal analysis was conducted for both the full model (see Figure 22.3) and one-quarter of the model under symmetrical conditions (see Figure 22.7). Table 22.2 lists the first five natural frequencies of the full model and one-quarter of the model. Obviously, they are quite different. Figures 22.8 through 22.12 plot the model shapes of the full model.

22.4.5 Discussion

Generally, sensitivity and operating frequency range are the two major parameters to evaluate accelerometer sensors. As sensitivity is defined as voltage output divided by acceleration, in combination with the relation between PZT thickness and voltage output, selecting an appropriate PZT

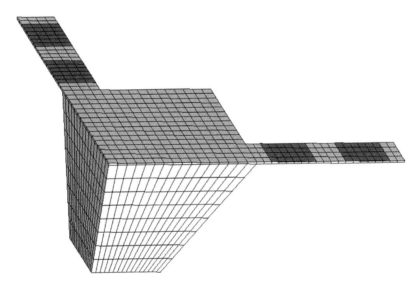

FIGURE 22.7
One-quarter of finite element model of the accelerator.

TABLE 22.2

The first five natural frequencies of one-quarter of the model and the full model

Natural frequency (Hz)	One-quarter of model	Full model
First	5594	7108
Second	27326	10624
Third	27369	24743
Fourth	73521	27392
Fifth	73574	27437

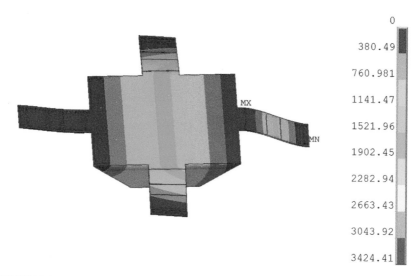

FIGURE 22.8
The first mode shape of the full model.

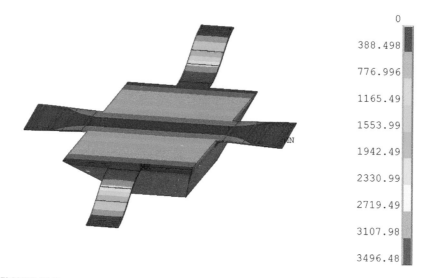

FIGURE 22.9
The second mode shape of the full model.

layer thickness can result in high sensitivity of the accelerometer. For the frequency range, compared to the natural frequencies of the full model, the one-quarter of the model misses some of the mode shapes, because PZT is anisotropic, especially since its piezoelectric constants are unsymmetrical. Therefore, it would be more appropriate to build a full model, although the structure of PZT is symmetrical.

Simulation of Piezoelectricity

FIGURE 22.10
The third mode shape of the full model.

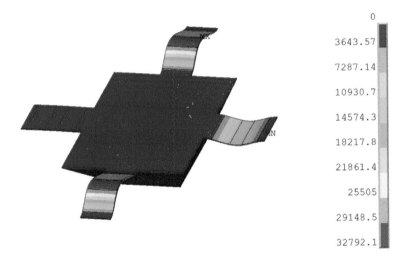

FIGURE 22.11
The fourth mode shape of the full model.

22.4.6 Summary

A three-dimensional finite element model of a piezoelectric thin film microaccelerometer was created, and the variation of the PZT layer thickness with the voltage output was studied. The obtained results can be used as guidelines for designing PZT thin-film microaccelerometers. Although the structure of PZT is symmetrical, it would be appropriate to build a full model because the material properties of PZT are unsymmetrical.

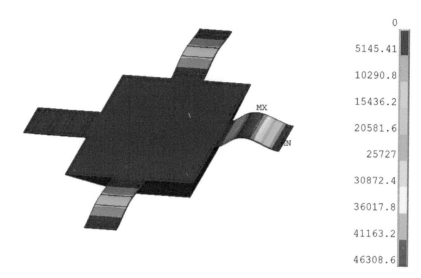

FIGURE 22.12
The fifth mode shape of the full model.

References

1. Curie, J., Curie, P., Développement, par pression, de l'électricité polaire dans les cristaux hémièdres faces 'inclinées, *Comptes Rendus de l'Academie des Sciences, Serie Generale*, Vol. 91, 1880, pp. 294–295.
2. Curie, J., Curie, P., Sur l'électricité polaire dans les cristaux hémièdres 'faces inclinées, *Comptes Rendus de l'Academie des Sciences, Serie Generale*, Vol. 91, 1880, pp. 383–386.
3. Dineva, P.S., Gross, D., Muller, R., Rangelov, T., *Dynamic Fracture of Piezoelectric Materials*, Springer, 2014.
4. ANSYS190 Help Documentation in the help page of product ANSYS190.
5. Gautschi, G., *Piezoelectric Sensorics: Force, Strain, Pressure, Acceleration and Acoustic Emission Sensors, Materials and Amplifiers*, Springer, 2002.
6. Uchino, K., *Piezoelectric Actuators and Ultrasonic Motors*, Kluwer, 1997.
7. Macdonald, G.A., A review of low-cost accelerometers for vehicle dynamics, *Sensors and Actuators A*, Vol. 21, 1990, pp. 303–307.
8. Wang, Q.-M., Yang, Z., Li, F., Smolinsk, P., Analysis of thin film piezoelectric microaccelerometer using analytical and finite element modeling, *Sensors and Actuators A*, Vol. 113, 2004, pp. 1–11.
9. Yu, J., Lan, C., System modeling of microaccelerometer using piezoelectric thin films, *Sensors and Actuators A*, Vol. 88, 2001, pp. 178–186.

23

Nanomaterials

Unlike the previously discussed materials, nanomaterials do not follow the traditional mechanics theory. Therefore, most of the material models in ANSYS cannot be applied to nanomaterials. However, the mechanical behaviors of nanomaterials can be studied in ANSYS when the nanomaterials are in the elastic stage. Chapter 23 introduces nanomaterials in Section 23.1 and determines the Young's modulus of Fe particles from experimental data by using finite element analysis in ANSYS in Section 23.2.

23.1 Introduction of Nano

The word "nano" comes from the Latin word *nanus* meaning dwarf. Therefore, a nanometer is one thousand millionth of a meter (1 nm = 1 × 10^{-9} m). A nanometer compares to a human hair like a human hair compares to a big house. Nanoscale materials refer to a group of substances with at least one dimension less than approximately 100 nm. Nanomaterials attract more and more interest because of the unique optical, magnetic, electrical, and mechanical properties that exist on that scale. These special properties may have a significant impact in electronics, medicine, and other fields. Some nanomaterials with different shapes, including gold, carbon, metals, meta-oxides, and alloys, are plotted in Figure 23.1.

Based on the modulation dimensionalities of nanomaterials, Siegel [1] classifies nanomaterials as follows: zero dimensional (atomic clusters), one dimensional (fibers, wires, and rods), two dimensional (films, plates, and networks), and three dimensional (consisting of equiaxed nanometer-sized grains) nanostructures.

The physical and chemical properties of nanomaterials at the nano-scale are much different from those at bulk size, which is due to the nanometer size of the materials resulting in a large fraction of surface atoms, high surface energy, spatial confinement, and reduced imperfections.

FIGURE 23.1
Nanoparticles (normalls ©123rf.com).

23.2 Determination of Young's Modulus of Fe Particles

23.2.1 Experiment

A compression test was performed on a single-crystal iron particle [1], which is a sphere with an outer diameter of 210 nm composed of Fe and enclosed by a 4-nm shell of $\gamma\text{-}Fe_2O_3$. In the test, the particles were compressed between a silicon substrate and a diamond tip at a constant displacement rate (see Figure 23.2a). During compression, the contact area increased with an increase in the external applied force until the particles broke suddenly. In order to verify the pure elastic deformation before the collapse, unloading was started prior to the loading reaching the critical point, returned to the starting point, and then reloaded. The force versus the compression displacement overlaps in the first and second loadings, which suggests that the particles underwent purely elastic deformation prior to the collapse. Consequently, the Young's modulus of Fe was determined from the experimental reaction forces using the finite element method.

23.2.2 Finite Element Model

The aforementioned experiment was simulated in ANSYS by using one-quarter of the finite element model (see Figure 23.2b), in which the sphere was in contact with the silicon substrate and the diamond tip. The silicon substrate and the diamond tip were meshed regularly using SOLID185, while the sphere was meshed freely using SOLID187. The contacts in the model were defined as standard.

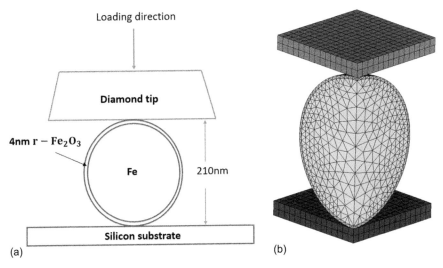

FIGURE 23.2
Experiment and its finite element model: (a) experiment, and (b) a quarter of the model.

23.2.3 Material Properties

The silicon, diamond, and γ-Fe_2O_3 were assumed to be linear elastic with Young's modulus values of 170 GPa, 1100 GPa, and 220 GPa, respectively [1].

23.2.4 Boundary Conditions and Loadings

The bottom of the silicon substrate was fixed with all degrees of freedom, and the top surface of the diamond tip was loaded with the displacement.

23.2.5 Solution

The goal of this study was to find the Young's modulus of a single-crystal iron by using the experimental data for reaction force. As the reaction force versus the displacement curve is nearly linear except at the starting stage, for simplification, the sum of the reaction forces in the last four substeps was selected for the objective. Initially, 20 GPa was assigned to the single-crystal iron. The whole calculation followed the flow chart shown in Figure 23.3. The experimental loading forces were interpolated from the experimental data.

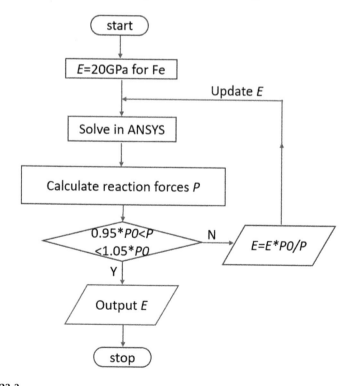

FIGURE 23.3
Flowchart to determine Young's modulus from the data on contact forces. P_0 is the experimental loading force.

23.2.6 Results

Computational results indicate that after two iterations, the obtained final reaction force reached the experimental data within 5% error. Then, a Young's modulus of 112.8 GPa defined in the final model was regarded as the material parameter of the single-crystal iron in the experiments. Figure 23.4 plots the computational results against the experimental data, showing that they match well.

23.2.7 Discussion

The scale of the single-crystal iron particles is nanometer; it is hard to measure the elastic parameters of the particles by using the conventional way. In this study, the Young's modulus of the single-crystal iron particles was obtained from the contact reaction forces by using the finite element method. The computational results significantly matching the experimental data validated the proposed method.

This indirect method is always adopted for nanoscale of mechanical behaviors in many fields. For example, to obtain the cell traction force, the displacement field caused by the cell traction force is calculated first by using the image process technology. Then, the cell traction force is computed from the displacement field inversely by using the finite element method [2].

The difference between the current finite element model and other models in the book is that the material parameter in the current finite element model

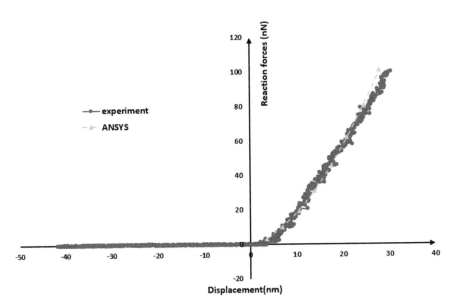

FIGURE 23.4
ANSYS results compared with the experimental data.

was updated based on the computational results. Therefore, the optimal process was formed within ANSYS. This method can be expanded for other applications such as optimization of mechanical design.

The geometry of the finite element model is on the nanometer scale ($1 \times ^{-9}$ m). To reduce the computational errors due to this scale, the following units were adopted in this study: mm for length, MPa for pressure, N for force.

23.2.8 Summary

Young's modulus of single-crystal iron particles was calculated from experimental data by using the finite element method. The reaction forces calculated using the defined Young's modulus were very close to the experimental data, which confirmed the validity of the obtained Young's modulus.

References

1. Han, W., Huang, L., Ogata, S., Kimizuka, H., Yang, Z., Weinberger, C., Li, Q., Liu, B., Zhang, X., Li, J., Ma, E., Shan, Z., From "smaller is stronger" to "size-independent strength plateau": towards measuring the ideal strength of iron, *Advanced Materials*, Vol. 27, 2015, pp. 3385–3390.
2. Yang, Z., Lin, J.S., Chen, J., Wang, J.H.C., Determining substrate displacement and cell traction fields–a new approach, *Journal of Theoretical Biology*, Vol. 242, 2006, pp. 607–616.

Part V

Retrospective

The previous four parts present the modeling of metals, polymers, soils, and modern materials. Based on this content, Part V summarizes material modeling, including association of material properties with the structure, temperature influence, solution setting, definition of material properties, and user subroutine.

24

Retrospective

The first four parts of this book focus on the structures and functions, as well as modeling of some conventional materials and modern materials. After reviewing these materials, the following conclusions can be drawn:

(1) Close association of material properties with the structure

In a metal, the atoms are bonded together by metallic bonds. The relative sliding of large groups of atoms causes the plastic deformation and ductility of the metal. On the other hand, macromolecules in a polymer are arranged in a network structure. The weak bonds between macromolecules give the polymer a low stiffness and high ductility. When the elongation of macromolecules approaches its limit, the forces within the macromolecules increase dramatically. This explains why polymers have strongly nonlinear behavior under large deformation. Soils are composed of rocks, water, and air. Water and air are in the voids of the soils between the rock particles. When soils are compacted, the rock particles get closer; because the stiffness of the soil increases with an increase in pore pressure, the stiffness of the soil rises. Consolidation of the soil occurs when water is expelled from the rocks. Different layers are assembled in composites, which make composites anisotropic. In functionally graded materials, the variation of the structure along the dimensions causes their properties to be functionally graded. Shape memory alloys under loadings or temperature change undergo phase transformation, which gives these alloys different mechanical behaviors. Poling makes ferroelectric materials have piezoelectricity. Overall, material properties are strongly related to the structure.

(2) Significant influence of temperature on material properties

Temperature affects materials in many aspects. First, it produces thermal strain because the material undergoes thermal expansion with a rise in temperature. Also, material properties are a function of temperature. Young's modulus and plastic parameters of metals always reduce with an increase in temperature. The Young's modulus of polymers is much different between glassy and rubbery regions. Finally, temperature change also affects the phase transformation of shape memory alloys.

(3) Various materials with different solution controls

Most plastic material models of metals and soils have a smooth yield surface. Therefore, the Newton–Raphson method is adopted in the nonlinear iteration, and the solution process exhibits very rapid convergence. However, the Mohr–Coulomb and Jointed Rock material models have non-smooth yield surfaces. Therefore, the cutting-plane algorithm using an elastic material tangent is adopted to solve the nonlinear equations for the Mohr–Coulomb and Jointed Rock materials. That requires selecting many substeps in the solution setting of the Mohr–Coulomb and Jointed Rock materials; these substeps show very slow convergence in the solution process.

(4) Various fields with different units

The determination of units in a model depends on the application field. For example, a study of such geomechanics problems as the tunnel excavation and slope stability analysis requires the selection of the units of force, length, and pressure as N, m, and Pa, respectively. When conducting medical research like the study of the breast shape after breast surgery, the units of force, length, and pressure – N, mm, and MPa – may be used. Trying to model MEMS and nanomaterials necessitates the adoption of μN, μm, and MPa for force, length, and pressure, respectively. Overall, for a different scale of problems the corresponding units for modeling should be selected.

(5) Anisotropic materials with symmetrical conditions

Normally, when the finite element model has symmetrical geometry and loading, it is always simplified to half a model with one direction of symmetry, one-quarter of a model with two directions of symmetry, and one-eighth of a model with three directions of symmetry to speed up the computation. However, in the case of anisotropic materials, such as piezoelectric materials and Jointed Rock materials, symmetrical conditions cannot be applied because the material properties are orientated, not isotropic.

(6) Application of four soil models

The Cam Clay model, Drucker–Prager model, Mohr–Coulomb model, and Jointed Rock model are four major soil models. They have different applications for the simulation of soils. For example, the Cam Clay model always works with a porous elastic model to simulate the subsidence of the ground. The Drucker–Prager model and the Mohr–Coulomb model only work with linear elastic materials to model rocks and concretes, although the extended Drucker–Prager model can work with different hardening models. The Jointed Rock model is used for modeling anisotropic rocks.

(7) Definition of material parameters

Material modeling requires the input of the material parameters for various materials. Normally, the material parameters can be found in manuals and references. If not, they can be obtained by the curve-fitting of experimental data. ANSYS provides curve-fitting functions for the Chaboche model, the creep models, and the hyperelastic models. In addition, the customers can calculate material parameters from the constitutive law. Some examples are given in Chapters 3, 9, and 21.

(8) User subroutine

ANSYS provides an extensive library of materials. However, when a material cannot be modeled in ANSYS, customers can develop user subroutines in ANSYS. Chapters 7 and 12 present two examples – UserCreep and UserHyper.

Appendix 1: Input File of Curve-Fitting of the Chaboche Model in Section 3.3

```
/COM, CURVE-FITTING OF THREE-TERM CHABOCHE CURVE-FITTING
WITH UNIAXIAL DATA
/COM, UNITS (N, MM, MPA)
*CREATE, LCF.EXP
 0.001264272    112.3015873
 0.001260138    -34.12698413
 0.001164631    -53.57142857
 0.001008229    -65.87301587
 0.000818357    -78.57142857
 0.000491711    -86.9047619
 0.000142897    -94.44444444
-0.000285125   -101.984127
-0.00071411    -107.5396825
-0.001140291   -110.7142857
-0.001303876   -112.3015873
-0.001295536    37.6984127
-0.001258384    46.82539683
-0.001190566    53.96825397
-0.001073243    61.11111111
-0.001009193    67.85714286
-0.000813625    76.98412698
-0.000553921    84.52380952
-9.199E-05      95.63492063
 0.000283722   100.7936508
 0.000643401   106.3492063
 0.000980473   109.5238095
 0.001153959   111.1111111
*END
/OUT, SCRATCH

/PREP7
TBFT, EADD, 1, UNIA, LCF.EXP
TBFT, FADD, 1, PLAS, CHAB, 3          ! THREE-TERM CHABOCHE
                                        MODEL

TBFT, SET, 1, PLAS, CHAB, 3, 1, 1E6
TBFT, SET, 1, PLAS, CHAB, 3, 2, 1E5
```

```
TBFT, SET, 1, PLAS, CHAB, 3, 3, 1E4
TBFT, SET, 1, PLAS, CHAB, 3, 4, 1E5
TBFT, SET, 1, PLAS, CHAB, 3, 5, 1E4
TBFT, SET, 1, PLAS, CHAB, 3, 6, 3000
TBFT, SET, 1, PLAS, CHAB, 3, 7, 55

TBFT, SOLVE, 1, PLAS, CHAB, 3, 0, 1000

/OUT
CFIT, , DFLAG, FITTEDDATLIST
CFIT, , CFNAME, PLCHAB3
CFIT, DEBUG

TBFT, LIST, 1

FINI
/EXIT
```

Appendix 2: Input File of the Forming Process Model in Section 4.2

```
/COM, UNITS (N, MM, MPA)
/PREP7

!************MATERIAL PROPERTIES**************
MP, EX, 1, 149650
MP, NUXY, 1, 0.33
TB, CHAB, 1, , 1          ! DEFINE CHABOCHE MATERIAL DATA
TBDATA, 1, 1.53E2, 62511, 1.1/311*62511
TB, RATE, 1, , 6, EVH     ! DEFINE RATE-DEPENDENT MATERIAL
                            DATA
TBDATA, 1, 1.53E2, 0, -1.53E2, 317, 1/7.7, 1150

K, 1, 49, 50
K, 2, 49, 6
K, 3, 44, 1
K, 4, 0, 1
K, 5, 44, 6
L, 2, 1
L, 4, 3
LARC, 3, 2, 5, 5,

K, 6, 100, 0
K, 7, 55, 0
K, 8, 50, -5
K, 9, 55, -5
LARC, 7, 8, 9, 5
K, 10, 50, -25
K, 11, 45, -30
K, 12, 45, -25
LARC, 10, 11, 12, 5
K, 13, 0, -30
L, 11, 13
L, 8, 10
L, 6, 7

RECTNG, 0, 95, 0, 1
ET, 1, 182
```

```
KEYOPT, 1, 3, 1
KEYOPT, 1, 6, 1
ESIZE, 0.5
AMESH, 1

ET, 2, 169
ET, 3, 172
R, 2, , , ,

TYPE, 2
REAL, 2
ESIZE, 0.25
LMESH, 1, 3
ALLSEL, ALL

TYPE, 3
REAL, 2
LSEL, S, LINE, , 11
NSLL, S, 1
ESLN, S
ESURF
ALLSEL

ET, 4, 169
ET, 5, 172
R, 3, , , ,

TYPE, 4
REAL, 3
ESIZE, 0.5
LMESH, 4, 8
ALLSEL, ALL

TYPE, 5
REAL, 3
LSEL, S, LINE, , 9
NSLL, S, 1
NSEL, R, LOC, X, 0, 95
ESLN, S
ESURF
ALLSEL

K, 21, 55, 1
K, 22, 100, 1
L, 21, 22
```

Appendix 2

```
ET, 6, 169
ET, 7, 172
R, 4, , , ,

TYPE, 6
REAL, 4
ESIZE, 0.5
LMESH, 13
ALLSEL, ALL

TYPE, 7
REAL, 4
LSEL, S, LINE, , 11
NSLL, S, 1
ESLN, S
ESURF
ALLSEL

ESEL, S, TYPE, , 1
NSLE, S
NSEL, R, LOC, X, 90, 95
D, ALL, UY
ESEL, S, TYPE, , 2
NSLE, S
D, ALL, UX, 0
D, ALL, UY, -30
ALLSEL
ESEL, S, TYPE, , 4, 6, 2
NSLE, S
D, ALL, ALL
ALLSEL

!--------------CREATE COMPONENTS--------------
ESEL, S, ENAME, , 182
CM, COMP1, ELEM              ! CREATE NLAD COMPONENT
ALLS
FINISH

/SOLU
!-----------GENERAL SOLUTION SETTINGS---------
NLGEOM, ON
RATE, ON
ERESX, NO
OUTRES, ALL, ALL
```

```
!----------NLAD SETTINGS------------      ! CHECK NLAD
                        PARAMETERS SECTION ABOVE FOR VALUES
NLAD, COMP1, ADD, MESH, SKEW, 0.95      ! NLAD WITH MESH-
                                        QUALITY-BASED CRITERION
NLAD, COMP1, ON, , , 1, 0, 1

!----------STEP SETTINGS----------
TIME, 1
NSUBST, 300, 10000, 100
NEQIT, 120
ALLS
SOLVE
FINI

/POST1
*GET, NSTP, ACTIVE, 0, SOLU, NCMSS
NSTP=103
*DIM, RFTOL, ARRAY, NSTP, 2
*DO, I, 1, NSTP
  ALLSEL
  SET, 1, I
    *GET, RF1, NODE, 574, RF, FY

  RFTOL(I, 2)=RF1
  *GET, RFTOL(I, 1), ACTIVE, 0, SET, TIME
*ENDDO
 *STATUS, RFTOL
```

Appendix 3: Input File of the Ratcheting Model in Section 5.2

```
/COM, UNITS (N, MM, MPA)
/PREP7
RECTNG, 0, 100, 0, 200
CYL4, 100, 0, 50
ASBA,    1,   2
ET, 1, 182
KEYOPT, 1, 3, 1
MSHAPE, 0, 2D
LESIZE, 3, , , 8, , , , , 1
LESIZE, 4, , , 10, , , , , 1
LESIZE, 10, , , 8, , , , , 1
LESIZE, 6, , , 10, , , , , 1
AMESH, 3

EP=5.47E4
NU=0.3
MP, EX, 1, EP
MP, NUXY, 1, NU
TB, NLISO, 1, , 2, 5        ! NONLINEAR ISOTROPIC HARDENING
POWER_N=0.1
SIGMA_Y=67.5
TBDATA, 1, SIGMA_Y, POWER_N
TB, CHAB, 1, , 3            ! CHABOCHE MODEL
TBDATA, 1, 67.5, 1E6, 9.37E4, 1E4, 1E5,
TBDATA, 6, 4.1E4, 1.1E3

NSEL, S, LOC, X, 0
D, ALL, UX
NSEL, S, LOC, Y, 0
D, ALL, UY
ALLSEL
FINISH

/SOLU
NSUBST, 50, 1000, 20
OUTRES, ALL, ALL
```

```
*DO, I, 1, 10                    ! CYCLIC LOADING
TIME, 5+(I-1)*8
NSEL, S, LOC, Y, 200
SF, ALL, PRES, -50
ALLSEL
SOLV
TIME, 8+(I-1)*8
NSEL, S, LOC, Y, 200
SF, ALL, PRES, 30
ALLSEL
SOLV
*ENDDO
FINISH
/POST26
ESOL, 2, 75, 4, S, Y,
ESOL, 3, 75, 4, EPEL, Y,
ESOL, 4, 75, 4, EPPL, Y,
ADD, 5, 3, 4
XVAR, 5
PLVAR, 2, , , , , , , , ,
```

Appendix 4: Input File of the Combustion Chamber Model in Section 6.2

```
/COM, UNITS (N, MM, MPA)
/PREP7
K, 1, -320, -535
K, 2, 0, -535
K, 3, 0, 0
K, 4, -320, 0

A, 1, 2, 3, 4

CYL4, -110, -265, 100
ASBA,    1,    2
CYL4, 0, 0, 80
ASBA,    3,    1
ET, 1, 182
KEYOPT, 1, 3, 2
KEYOPT, 1, 1, 1

ESIZE, 25
AMESH, ALL

ALLSEL
FLST, 5, 13, 2, ORDE, 8
FITEM, 5, 14
FITEM, 5, -18
FITEM, 5, 137
FITEM, 5, -138
FITEM, 5, 147
FITEM, 5, -148
FITEM, 5, 159
FITEM, 5, -162
ESEL, S, , , P51X
FLST, 5, 13, 2, ORDE, 8
FITEM, 5, 14
FITEM, 5, -18
FITEM, 5, 137
FITEM, 5, -138
FITEM, 5, 147
```

```
FITEM, 5, -148
FITEM, 5, 159
FITEM, 5, -162
CM, _Y, ELEM
ESEL, , , , P51X
CM, _Y1, ELEM
CMSEL, S, _Y
CMDELE, _Y
EREF, _Y1, , , 1, 0, 1, 1
CMDELE, _Y1
ALLSEL
EPLOT


TB, ELAS, 1, 3                  ! ELASTIC PROPERTIES
TBTEMP, 20
TBDATA, 1, 76000, 0.33
TBTEMP, 100
TBDATA, 1, 72000, 0.33
TBTEMP, 300
TBDATA, 1, 60000, 0.33

TB, CHAB, 1, 4, 1               ! CHABOCHE MODEL
TBTEMP, 20
TBDATA, 1, 80, 21700, 155
TBTEMP, 100
TBDATA, 1, 63, 22000, 200
TBTEMP, 200
TBDATA, 1, 52, 8460, 310
TBTEMP, 300
TBDATA, 1, 4, 3780, 420

TB, RATE, 1, 3, , PERZYNA       ! RATE TABLE
TBTEMP, 20
TBDATA, 1, 1/8.5, 1E-4
TBTEMP, 200
TBDATA, 1, 1/8.5, 3E-6
TBTEMP, 300
TBDATA, 1, 1/8.5, 1E-10

MP, ALPX, 1, 6E-6

TB, PLAS, 1, 4, , MISO          ! MULTILINEAR ISOTROPIC
                                  HARDENING
TBTEMP, 20
```

Appendix 4

```
TBPT, DEFI, 0, 80
TBPT, DEFI, 0.1, 200
TBTEMP, 100
TBPT, DEFI, 0, 63
TBPT, DEFI, 0.1, 157
TBTEMP, 200
TBPT, DEFI, 0, 52
TBPT, DEFI, 0.1, 130
TBTEMP, 300
TBPT, DEFI, 0, 4
TBPT, DEFI, 0.1, 10

NSEL, S, LOC, X, 0
NSEL, A, LOC, X, -320
D, ALL, UX, 0
NSEL, S, LOC, Y, 0
NSEL, A, LOC, Y, -535
D, ALL, UY, 0
ALLSEL

TREF, 0
FINISH

/SOLU
NSUBST, 100, 1000, 30
KBC, 0
RATE, ON
OUTRES, ALL, ALL
TIME, 80
BF, ALL, TEMP, 280
SOLVE

TIME, 160
BF, ALL, TEMP, 80
SOLVE

TIME, 240
BF, ALL, TEMP, 280
SOLVE

TIME, 320
BF, ALL, TEMP, 80
SOLVE
```

```
TIME, 400
BF, ALL, TEMP, 280
SOLVE

TIME, 500
BF, ALL, TEMP, 80
SOLVE

FINISH

/POST26
ESOL, 2, 288, 290, EPPL, X, EPPLEQV
ESOL, 3, 288, 290, S, EQV,
ESOL, 4, 288, 290, EPTH, X,
PLVAR, 2
PLVAR, 3
PLVAR, 4
```

Appendix 5: Input File of the Bolt Model under Pretension in Section 7.2

```
/COM, UNITS (LBF, INCH, PSI)
/UPF, USERCREEP.F
/PREP7
ET, 1, 187
MP, EX, 1, 14.665E6
MP, PRXY, 1, 0.30
TB, PLAS, 1, 1, 2, MISO          ! MULTILINEAR ISOTROPIC
                                   HARDENING
TBPT, DEFI, 0, 29.33E3
TBPT, DEFI, 1.59E-3, 50E3

MP, EX, 2, 42E6
MP, PRXY, 2, 0.30
TB, PLAS, 2, 1, 2, MISO          ! MULTILINEAR ISOTROPIC
                                   HARDENING
TBPT, DEFI, 0, 29.33E3*3
TBPT, DEFI, 1.59E-3, 50E3*3
TB, CREEP, 2, 1, 4, 100          ! USER CREEP
TBDATA, 1, 0.54E-16, 2.3, -0.6

CYLIND, 0.5, , -0.25, 0, 0, 180
CYLIND, 0.5, , 1, 1.25, 0, 180
CYLIND, 0.25, , 0, 1, 0, 180
WPOFF, .05
CYLIND, 0.35, 1, 0, 0.75, 0, 180
WPOFF, -.1
CYLIND, 0.35, 1, 0.75, 1, 0, 180
WPSTYLE, , , , , , , 0
VGLUE, ALL
NUMC, ALL
MAT, 1
SMRT, OFF
VMESH, 4, 5
MAT, 2
VMESH, 1, 3
PSMESH, , BOLT, , VOLU, 1, 0, Z, 0.5, , , , ELEMS
FINISH
```

```
/SOLU
TIME, 1E-4
KBC, 1
RATE, 0
NSEL, S, LOC, Y, 0
D, ALL, UY, 0
ALLSEL
D, NODE(0, 0, -0.25), UX, 0
D, NODE(0, 0, -0.25), UZ, 0
D, NODE(0, 0, 1.25), UX, 0
RESCONTROL, LINEAR, ALL, 1
SLOAD, 1, PL01, TINY, DISP, 0.01, 1, 2
SOLVE
TIME, 1.5E6
RATE, 1
NSUBST, 100
OUTRES, ALL, ALL
SOLVE
```

Appendix 6: Input File of Curve-Fitting of the Ogden Model in Section 9.3

```
*CREATE, CFHY-OG-U.EXP    ! INPUT THE EXPERIMENTAL DATA
! ENGINEERING STRAIN        ENGINEERING STRESS (MPA)
0.265384615               0.2125
0.403846154               0.3125
0.588461538               0.4
0.853846154               0.475
1.153846154               0.6
1.396153846               0.675
1.996153846               0.8625
2.55                      1.0375
2.988461538               1.225
3.738461538               1.5625
4.303846154               1.9375
4.707692308               2.3
5.076923077               2.675
5.342307692               3
5.538461538               3.3625
5.792307692               3.7125
6.011538462               4.0625
6.092307692               4.45
6.173076923               4.7875
6.334615385               5.1375
6.403846154               5.5375
*END

*CREATE, CFHY-OG-B.EXP    ! INPUT THE EXPERIMENTAL DATA
! ENGINEERING STRAIN        ENGINEERING STRESS (MPA)

0.173076923               0.3
0.288461538               0.425
0.438461538               0.5125
0.692307692               0.6375
0.934615385               0.75
1.465384615               0.975
1.996153846               1.225
2.4                       1.4375
2.734615385               1.6625
```

```
    3.034615385     1.925
    3.219230769     2.1875
    3.403846154     2.4
*END

/PREP7
! DEFINE UNIAXIAL DATA
TBFT, EADD, 1, UNIA, CFHY-OG-U.EXP
TBFT, EADD, 1, BIAX, CFHY-OG-B.EXP
! DEFINE MATERIAL: TWO-PARAMETER OGDEN MODEL
TBFT, FADD, 1, HYPER, OGDEN, 2
! SET THE INITIAL VALUE OF COEFFICIENTS
TBFT, SET, 1, HYPER, OGDE, 2, 1, 1
TBFT, SET, 1, HYPER, OGDE, 2, 2, 1
TBFT, SET, 1, HYPER, OGDE, 2, 3, 1
TBFT, SET, 1, HYPER, OGDE, 2, 4, 1

! DEFINE SOLUTION PARAMETERS IF ANY
TBFT, SOLVE, 1, HYPER, OGDEN, 2 , 1, 500

/OUT
! PRINT THE RESULTS
TBFT, LIST, 1
TBFT, FSET, 1, HYPER, OGDEN, 2
TBLIS, ALL, ALL

/OUT
CFIT, , DFLAG, FITTEDDATLIST
CFIT, DEBUG
FINISH
```

Appendix 7: Input File of the Rubber Rod Model under Compression in Section 9.4

```
/COM, UNITS (N, M, PA)
/PREP7
ET, 1, 182
KEYOPT, 1, 3, 2
KEYOPT, 1, 1, 0
! MATERIAL PROPERTIES
TB, HYPER, 1, , 3, MOONEY            !ACTIVATE THREE-TERM
                                     MOONEY-RIVLIN
TBDATA, 1, 1.89E5, -1.933E3, 1.183E2

!TB, HYPER, 1, , 1, BOYCE            ! ARRUDA-BOYCE MODEL
!TBDATA, 1, 3.148E5, 5.032

!TB, HYPER, 1, , 2, OGDEN            ! TWO-TERM OGDEN
!TBDATA, 1, 3.565E4, 3.398, 1.272E9, 4.948E-4

! CREATE GEOMETRY OF TIRE
K, 1, -1.5, 0.05
K, 2, -1.5, -1
K, 3, 1.5, -1
K, 4, 1.5, 0.05
L, 2, 1
L, 3, 2
L, 4, 3
K, 5, -2, 1
K, 6, 2, 1
L, 5, 6
CYL4, 0, 0, 1
RECTNG, -0.25, 0.25, -0.25, 0.25
AOVLAP, ALL
ESIZE, 0.08
MSHAPE, 0, 2D
AMESH, 2
AMESH, 3
ALLSEL
```

```
ET, 2, 169
ET, 3, 172
KEYOPT, 3, 9, 1
R, 2, , , ,

TYPE, 2
REAL, 2
ESIZE, 0.2
LMESH, 4
ALLSEL, ALL

TYPE, 3
REAL, 2
LSEL, S, LINE, , 5, 6
NSLL, S, 1
ESLN, S
ESURF
ALLSEL

ET, 4, 169
ET, 5, 172
KEYOPT, 5, 9, 1
R, 3, , , ,

TYPE, 4
REAL, 3
ESIZE, 0.2
LMESH, 1, 3
ALLSEL, ALL

TYPE, 5
REAL, 3
LSEL, S, LINE, , 5, 8
NSLL, S, 1
ESLN, S
ESURF
ALLSEL

LSEL, S, LINE, , 1, 3
NSLL, S, 1
D, ALL, ALL
ALLSEL

LSEL, S, LINE, , 4
NSLL, S, 1
```

Appendix 7

```
D, ALL, UX, 0
D, ALL, UY, -0.95

ALLSEL
FINISH

/SOLV
ANTYPE, 0
NLGEOM, ON

TIME, 1
OUTRES, ALL, ALL
NSUBST, 50, 5000, 10
KBC, 0
SOLVE
```

Appendix 8: Input File of the Liver Soft Tissue Model in Section 10.4

```
/COM, UNITS (N, MM, MPA)
/PREP7
K,  1, -3.50,     0.00,    0.00
K,  2, -2.00,     1.50,    0.00
K,  3, -0.5,      2.0,     0.00
K,  4,  1.50,     1.50,    0.00
K,  5,  3.00,     0.00,    0.00
K,  6, -5.00,     0.00,    0.00
K,  7,  5.00,     0.00,    0.00
K,  8,  5.00,    -0.50,    0.00
K,  9, -5.00,    -0.50,    0.00
K, 10, -0.45151,  3
BSPLIN, 1, 2, 3, 4, 5
L, 1, 5
AL, 1, 2
A, 6, 7, 8, 9
ET, 1, 182
KEYOPT, 1, 3, 2
ESIZE, 0.2
AMESH, ALL
! ********************* MATERIAL CONSTANTS **********
TB, HYPE, 1, 1, 1, OGDEN          ! OGDEN MODEL
TBDATA, 1, 3.7E-2, 5.59,
TB, PRONY, 1, 1, 3, SHEAR         ! PRONY'S SERIES
TBDATA, 1, 6.9701E-3, 10, 5.83E-2, 100, 3.53E-2, 1000

MP, EX, 5, 1E6
MP, NUXY, 5, 0.2

MP, MU, 2, 0.3                    ! CONTACT
MAT, 2
R, 3
REAL, 3
ET, 2, 169
ET, 3, 172
KEYOPT, 3, 9, 0
KEYOPT, 3, 10, 0
```

```
R, 3,
RMORE,
RMORE, , 0
RMORE, 0
! GENERATE THE TARGET SURFACE
LSEL, S, , , 3
CM, _TARGET, LINE
TYPE, 2
NSLL, S, 1
ESLN, S, 0
ESURF
CMSEL, S, _ELEMCM
! GENERATE THE CONTACT SURFACE
LSEL, S, , , 2
CM, _CONTACT, LINE
TYPE, 3
NSLL, S, 1
ESLN, S, 0
ESURF
ALLSEL
NSEL, S, LOC, Y, -0.5
D, ALL, UY, 0
NSEL, R, LOC, X, -5
D, ALL, ALL
ALLSEL

CYL4, -0.45151, 2.5, 0.5
ESIZE, 0.1
AMESH, 3

MP, MU, 3, 0
MAT, 3
MP, EMIS, 1, 7.88860905221E-031
R, 5
REAL, 5
ET, 6, 169
ET, 7, 172
R, 5, , , 1.0, 0.1, 0,
RMORE, , , 1.0E20, 0.0, 1.0,
RMORE, 0.0, 0, 1.0, , 1.0, 0.5
RMORE, 0, 1.0, 1.0, 0.0, , 1.0
RMORE, , , , , 1.0
KEYOPT, 7, 3, 0
KEYOPT, 7, 4, 0
KEYOPT, 7, 5, 3
```

Appendix 8

```
KEYOPT, 7, 7, 0
KEYOPT, 7, 8, 0
KEYOPT, 7, 9, 1
KEYOPT, 7, 10, 0
KEYOPT, 7, 11, 0
KEYOPT, 7, 12, 0
KEYOPT, 7, 14, 0
KEYOPT, 7, 18, 0
KEYOPT, 7, 2, 0
! GENERATE THE TARGET SURFACE
LSEL, S, , , 9
LSEL, A, , , 10
CM, _TARGET, LINE
TYPE, 6
NSLL, S, 1
ESLN, S, 0
ESURF
CMSEL, S, _ELEMCM
! GENERATE THE CONTACT SURFACE
LSEL, S, , , 1
CM, _CONTACT, LINE
TYPE, 7
NSLL, S, 1
ESLN, S, 0
ESURF
ALLSEL

ET, 8, 169
ET, 9, 172
KEYOPT, 9, 2, 2
KEYOPT, 9, 12, 5
R, 9

! PILOT NODE
TSHAP, PILOT
N, 9999, -0.45151, 2.5
TYPE, 8
REAL, 9
E, 9999

TYPE, 9
REAL, 9
ASEL, S, AREA, , 3
NSLA, S, 1
ESLN, S
```

```
ESURF
ALLSEL, ALL
D, 57, UX
ALLSEL

ASEL, S, AREA, , 2, 3
ESLA, S
EMODIF, ALL, MAT, 5
ALLSEL
FINISH
/SOLU
TIME, 0.1
NLGEOM, ON
NSUBST, 50, 1000, 30
KBC, 0
D, 9999, UY, -0.5
D, 9999, UX, 0, , , , ROTZ
ALLSEL
OUTRES, ALL, ALL
SOLVE
TIME, 1000
SOLVE
FINISH

/POST1
*GET, NSTP, ACTIVE, 0, SOLU, NCMSS
*DIM, RFTOL, ARRAY, NSTP, 2
*DO, I, 1, NSTP
    ALLSEL
    SET, 1, I
    NSEL, S, LOC, Y, -0.5
    *GET, NUM_N, NODE, 0, COUNT
    *GET, N_MIN, NODE, 0, NUM, MIN
    RFT=0
     *DO, J, 1, NUM_N, 1
        CURR_N=N_MIN
         *GET, RF1, NODE, CURR_N, RF, FY
       RFT=RFT+RF1
         *GET, N_MIN, NODE, CURR_N, NXTH
     *ENDDO
    RFTOL(I, 2)=RFT
    *GET, RFTOL(I, 1), ACTIVE, 0, SET, TIME
*ENDDO
*STATUS, RFTOL
```

Appendix 9: Input File of the Rubber Tire Damage Model in Section 11.3

```
/COM, UNITS (N, M, Pa)
/PREP7
R1=2.0            ! EXTERIOR RADIUS OF TIRE, METER
R2=1.0            ! INTERIOR RADIUS OF TIRE
TK=1.0            ! THICKNESS
L=4.0             ! LENGTH FOR ROAD
PI=3.1416

ET, 1, SOLID185

! MATERIAL PROPERTIES (YEOH)
TB, HYPER, 1, , 3, YEOH       !ACTIVATE 3 TERM YEOH DATA
                               TABLE
TBDATA, 1, 1.317e6       !Define C1
TBDATA, 2, 0.0341e6      !Define C2
TBDATA, 3, 0.0037e6      !Define C3
TBDATA, 4, 3.0E-9        !Define first incompressibility
                          parameter

TB, CDM, 1, , 3, PSE2    ! modified Ogden-Roxburgh
                           Mullins effect
TBDATA, 1, 2.34, 0.069e6, 0.222
! CREATE GEOMETRY OF TIRE
CYL4, 0, 0, R2, 0, R1, 180, TK
LESIZE, 5, , , 45
LESIZE, 6, , , 4
LESIZE, 11, , , 2
MSHAPE, 0, 3D
MSHKEY, 1
VMESH, ALL
VSYMM, Y, ALL
ALLSEL
NUMMRG, NODE
ALLSEL

! DEFINE COUPLING OF CENTRAL POINT WITH INTERNAL SURFACE
ET, 2, 170
```

```
ET, 3, 174
KEYOPT, 3, 2, 2
KEYOPT, 3, 12, 5
R, 3

! PILOT NODE
TSHAP, PILOT
N, 9999, 0, 0, 0        ! CENTER OF TIRE
TYPE, 2
REAL, 3
E, 9999

TYPE, 3
REAL, 3
ASEL, S, AREA, , 4, 10, 6
NSLA, S, 1
ESLN, S
ESURF
ALLSEL, ALL

! CREATE ROAD SURFACE
K, 101, -L/2, -R1, -L/2+TK
K, 102, L/2, -R1, -L/2+TK
K, 103, L/2, -R1, L/2
K, 104, -L/2, -R1, L/2
A, 104, 103, 102, 101


ET, 4, 170
ET, 5, 174
R, 4, , , ,

TYPE, 4
REAL, 4
ESIZE, 3
AMESH, 13
ALLSEL, ALL

TYPE, 5
REAL, 4
ASEL, S, AREA, , 3, 9, 6
NSLA, S, 1
ESLN, S
ESURF
ALLSEL
```

Appendix 9

```
! BOUNDARY CONDITIONS
D, 9999, UX, 0, , , , UY, UZ, ROTX, ROTY
D, 9999, ROTZ, 0
ALLSEL

FINISH

/SOLV
ANTYPE, 0
NLGEOM, ON

TIME, 1
OUTRES, ALL, ALL
NSUBST, 10, 1000, 10
KBC, 0

ASEL, S, AREA, , 13
NSLA, S, 1
D, ALL, UX, 0, , , , UZ, ROTX, ROTY, ROTZ
D, ALL, UY, 0.2
ALLSEL
SOLVE

TIME, 10
OUTRES, ALL, ALL
NSUBST, 100, 1000, 100
KBC, 0

D, 9999, UX, 0, , , , UY, UZ, ROTX, ROTY
D, 9999, ROTZ, 6*PI
ALLSEL
SOLVE
FINISH

/POST1
NN=110
*DIM, FYY, ARRAY, NN
*DIM, TT, ARRAY, NN
*DIM, MZZ, ARRAY, NN
*DO, I, 1, 10
   SET, 1, I
   *GET, FYY(I), NODE, 9999, RF, FY
   *GET, TT(I), ACTIVE, 0, SET, TIME
   *GET, MZZ(I), NODE, 9999, RF, MZ
```

```
*ENDDO
*DO, I, 1, 100
   II=I+10
   SET, 2, I
   *GET, FYY(II), NODE, 9999, RF, FY
   *GET, TT(II), ACTIVE, 0, SET, TIME
   *GET, MZZ(II), NODE, 9999, RF, MZ
*ENDDO
```

Appendix 10: Input File of UserHyper in Section 12.2

```
/COM, UNITS (N, MM, MPA)
/PREP7
! USE USER MATERIAL SUBROUTINE
/UPF, USERHYPER.F
ET, 1, 185
MAT, 1
BLOCK, 0, 1, 0, 1, 0, 1
ESIZE, 0.5
VMESH, 1
! DEFINE MATERIAL BY TB, USER
TB, HYPER, 1, , 3, USER
TBDATA, 1, 1.5, 10, 1E-3
NSEL, S, LOC, X
D, ALL, UX, 0
NSEL, ALL
NSEL, S, LOC, Y
D, ALL, UY, 0
NSEL, ALL
NSEL, S, LOC, Z
D, ALL, UZ, 0
ALLSEL, ALL
/SOLU
ANTYPE, STATIC
NLGEOM, ON
NSEL, S, LOC, Y, 1
D, ALL, UY, 2.4
ALLSEL, ALL
NSUBS, 20, 1000, 10
TIME, 1
OUTRES, ALL, ALL
OUTPR, ALL, ALL
SOLV
FINI
```

```
/POST26
NUMVAR, 200
ESOL, 2, 7, 15, S, Y,
/SHOW, 'USERMAT-185', 'GRPH', ' '
PLVAR, 2, ,
PRVAR, 2
```

Appendix 11: Input File of the Tower Subsidence Model in Section 14.3

```
/COM, UNITS (N, M, PA)
/PREP7
TB, PELAS, 1, , , POISSON              ! POROELASTIC
                                          MATERIAL
TBDATA, 1, 0.026, 5.0E3, 0.28, 0.889
TB, SOIL, 1, , , CAMCLAY               ! CAM CLAY MODEL
TBDATA, 1, 0.1174, 1, 103E3, 103E3, 1
TBDATA, 6, 1, 1
MP, DENS, 1, 1923

MP, EX, 2, 2E10
MP, NUXY, 2, 0.3
MP, DENS, 2, 5000

K, 1
K, 2, 30
K, 3, 30, 30
K, 4, 0, 30
K, 5, 0, 20
K, 6, 10, 20
K, 7, 10, 30
K, 8, 8, 30
K, 9, 5, 36
K, 10, 0, 36
K, 11, 0, 30
A, 1, 2, 3, 4
A, 5, 6, 7, 4
A, 8, 9, 10, 11
AOVLAP, 1, 2

ET, 1, 182
KEYOPT, 1, 3, 1

TYPE, 1
MAT, 1
ESIZE, 1
```

```
MSHAPE, 0, 2D
AMESH, 2
AMESH, 4

TYPE, 1
MAT, 2
AMESH, 3
ALLSEL

R, 3
REAL, 3
ET, 2, 169
ET, 3, 172
KEYOPT, 3, 9, 1
KEYOPT, 3, 10, 0
R, 3,
RMORE,
RMORE, , 0
RMORE, 0
! GENERATE THE TARGET SURFACE
LSEL, S, , , 12
CM, _TARGET, LINE
TYPE, 2
NSLL, S, 1
ESLN, S, 0
ESURF
CMSEL, S, _ELEMCM
! GENERATE THE CONTACT SURFACE
LSEL, S, , , 7
CM, _CONTACT, LINE
TYPE, 3
NSLL, S, 1
ESLN, S, 0
ESURF
ALLSEL

! BOUNDARY CONDITIONS
NSEL, S, LOC, X, 0
D, ALL, UX, 0
NSEL, S, LOC, Y, 0
D, ALL, UY, 0
NSEL, S, LOC, X, 30
```

Appendix 11

```
D, ALL, UX, 0
ALLSEL
FINISH

/SOLU
ACEL, , 9.81            ! SET STANDARD GRAVITY FOR DEAD
LOAD
ALLSEL
ANTYPE, STATIC, NEW
NLGEOM, ON
NSUBST, 100, 5000, 50
TIME, 1
OUTRES, ALL, ALL
OUTRES, SVAR, ALL
ALLS
SOLVE
FINI
```

Appendix 12: Input File of the Soil–Arch Interaction Model in Section 15.2

```
/COM, UNITS (N, MM, MPA)
/PREP7
PCIRC, 2090, 1875, 36.87, 143.13,
K, 5, 1670+1500, 1250, 0
K, 6, -1670-1500, 1250, 0
K, 7, 1670+1500, 3090, 0
K, 8, -1670-1500, 3090, 0
L, 1, 5
L, 5, 7
L, 7, 8
L, 8, 6
L, 6, 2
LSEL, S, LINE, , 5, 9
LSEL, A, LINE, , 1
AL, ALL
ET, 1, 182
KEYOPT, 1, 3, 2
MSHAPE, 0, 2D
ESIZE, 125
AMESH, 2
ADELE, 1
PCIRC, 2090, 1875, 36.87, 143.13,
ET, 2, 182
KEYOPT, 1, 3, 2
ESIZE, 125
TYPE, 2
AMESH, 1
/GSAV, CWZ, GSAV, , TEMP
MP, MU, 1, 0.7
MAT, 1
R, 3
REAL, 3
ET, 3, 169
ET, 4, 172
KEYOPT, 4, 9, 0
KEYOPT, 4, 10, 0
R, 3,
```

```
RMORE,
RMORE, , 0
RMORE, 0
! GENERATE THE TARGET SURFACE
LSEL, S, , , 10
CM, _TARGET, LINE
TYPE, 3
NSLL, S, 1
ESLN, S, 0
ESURF
CMSEL, S, _ELEMCM
! GENERATE THE CONTACT SURFACE
LSEL, S, , , 1
CM, _CONTACT, LINE
TYPE, 4
NSLL, S, 1
ESLN, S, 0
ESURF
ALLSEL

MP, EX, 2, 16000
MP, NUXY, 2, 0.2

MP, EX, 3, 1000
MP, NUXY, 3, 0.2
! SET NONLINEAR PARAMETERS FOR INTACT ROCK
BETA=1.05                  ! DRUCKER-PRAGER PARAMETER
                             (SLOPE OF CONE)
SIGMA_Y=0.0224             ! DRUCKER-PRAGER PARAMETER
                             (STRENGTH)
TB, EDP, 3, 1, 2, LYFUN    ! EXTENDED DRUCKER-PRAGER
                             MODEL
TBDATA, 1, BETA, SIGMA_Y

TB, EDP, 3, 1, 2, LFPOT    ! ASSOCIATED PLASTIC FLOW
TBDATA, 1, BETA

ESEL, S, TYPE, , 1
EMODIF, ALL, MAT, 3

ESEL, S, TYPE, , 2
EMODIF, ALL, MAT, 2
ALLSEL
FINISH
```

Appendix 12

```
/SOLU
LSEL, S, LINE, , 6, 8, 2
NSLL, S, 1
D, ALL, UX, 0
LSEL, S, LINE, , 5, 9, 4
NSLL, S, 1
D, ALL, UY, 0
LSEL, S, LINE, , 11, 13, 2
NSLL, S, 1
D, ALL, ALL
ALLSEL
NSEL, S, NODE, , 97, 99
SF, ALL, PRES, 0.3
ALLSEL
NSUBST, 40, 1000, 20
OUTRES, ALL, ALL
SOLV
```

Appendix 13: Input File of the Slope Stability Model in Section 16.3

```
/COM, UNITS (N, M, PA)
/PREP7
ET, 1, 182
KEYOPT, 1, 3, 2
KEYOPT, 1, 1, 2
K, 1, 0, 0
K, 2, 15, 0
K, 3, 30, 0
K, 4, 40, 0
K, 5, 40, 10
K, 6, 30, 10
K, 7, 15, 10
K, 8, 15, 20
K, 9, 0, 20
K, 10, 0, 10
A, 1, 2, 7, 10
A, 10, 7, 8, 9
A, 2, 3, 6, 7
A, 3, 4, 5, 6
A, 7, 6, 8
AGLUE, ALL
MSHKEY, 1
ESIZE, 1
AMESH, ALL
ALLSEL
MP, EX, 1, 2.8E10
MP, NUXY, 1, 0.2
MP, DENS, 1, 2500
TB, MC, 1, , , BASE                              ! MOHR-
                                                   COULOMB MODEL
TBDATA, 1, 15.96, 9.20E3, 0, 15.96, 9.20E3   ! FT = 1.63
TB, MC, 1, , , RCUT
TBDATA, 1, 0.5E3, 0.5E3
TB, MC, 1, , , MSOL
TBDATA, 1, 2
```

```
ALLSEL
NSEL, S, LOC, X, 0
D, ALL, UX, 0
NSEL, S, LOC, Y, 0
D, ALL, UY, 0
NSEL, S, LOC, X, 40
D, ALL, UX, 0
ALLSEL
ACEL, , 9.81              ! SET STANDARD GRAVITY FOR
                            DEAD LOAD

ALLSEL
FINISH

/SOLV
ANTYPE, STATIC, NEW
NLGEOM, ON
NSUBST, 100, 5000, 50
TIME, 1
OUTRES, ALL, ALL
OUTRES, SVAR, ALL
ALLS
SOLVE
FINI
```

Appendix 14: Input File of the Tunnel Excavation Model in Section 17.3

```
/COM, UNITS (N, M, PA)
/PREP7
RECTNG, -5, 5, -5, 5
CYL4, 0, 0, 0, 0, 0.75, 360
ASBA,    1,    2, , , KEEP
ET, 1, 182
KEYOPT, 1, 3, 2
ESIZE, 0.25
LESIZE, 5, , , 10
MSHAPE, 0, 2D
AMESH, ALL
ALLSEL

MP, EX, 1, 200E6
MP, NUXY, 1, 0.23
MP, DENS, 1, 1810
TB, JROCK, 1, , , BASE                  ! BASE MATERIAL
                                          FOR JOINTED ROCK
TBDATA, 1, 21.2, 1E6, 10, 21.2, 1E6
TB, JROCK, 1, , , RCUT
TBDATA, 1, 0.1E6, 0.1E6

TB, JROCK, 1, , , RSC
TBDATA, 1, 0

TB, JROCK, 1, , , FPLANE                ! FAILURE PLANE
TBDATA, 1, 15.8, 0.1E6, 0, 15.8, 0.1E6

TB, JROCK, 1, , , FTCUT
TBDATA, 1, 0.05E6, 0.05E6

TB, JROCK, 1, , , FORIE
TBDATA, 1, , -60

NSEL, S, LOC, X, -5
D, ALL, UX, 0
NSEL, S, LOC, Y, -5
```

```
D, ALL, UY, 0
ALLSEL
NSEL, S, LOC, X, 5
D, ALL, UX, 0
NSEL, S, LOC, Y, 5
D, ALL, UY, 0
ALLSEL

INISTATE, SET, DTYP, STRE
INISTATE, DEFI, ALL, , , , -1E6, -1E6, , 0, 0, 0

ALLSEL
FINISH

FINISH

/SOLV
ANTYPE, STATIC, NEW
TIME, 1
NLGEOM, ON
NSUBST, 1
SOLVE
TIME, 2
ASEL, S, AREA, , 2
ESLA, S
EKILL, ALL                      ! EKILL COMMAND
ALLSEL
OUTRES, ALL, ALL
NSUBST, 10, 1000, 10
SOLVE
FINI
```

Appendix 15: Input File of the Settlement Model in Section 18.3

```
/COM, UNITS (N, M, PA)
/PREP7
ET, 1, 182
KEYOPT, 1, 3, 2
ET, 2, 212
KEYOPT, 2, 3, 2
KEYOPT, 2, 12, 1

K, 1
K, 2, 5
K, 3, 395
K, 4, 400
K, 5, 395, 50
K, 6, 395, 100
K, 7, 400, 125
K, 8, 395, 125
K, 9, 5, 125
K, 10, 0, 125
K, 11, 5, 75
K, 12, 5, 25

K, 13, 0, 25
K, 14, 0, 75
K, 15, 400, 50
K, 16, 400, 100

A, 1, 2, 12, 13
A, 3, 4, 15, 5
A, 2, 3, 5, 12
A, 5, 6, 11, 12
A, 6, 8, 9, 11

CYL4, 100, 58, 0.2
CYL4, 200, 63, 0.2
CYL4, 300, 68, 0.2
```

```
ASBA, 4, 6
ASBA, 9, 7
ASBA, 4, 8

A, 12, 11, 14, 13
A, 11, 9, 10, 14
A, 15, 16, 6, 5
A, 16, 7, 8, 6

LSEL, S, LINE, , 17, 28
LESIZE, ALL, , , 8, , , , , 1
LSEL, S, LINE, , 10, 12, 2
LESIZE, ALL, , , 80, , , , , 1
LSEL, S, LINE, , 11, 13, 2
LESIZE, ALL, , , 12, , , , , 1
ALLSEL
TYPE, 2
MSHAPE, 0, 2D
MSHKEY, 0
MAT, 2
AMESH, 6

MAT, 1
TYPE, 1
MSHKEY, 1
AMESH, 3
AMESH, 5

ALLSEL
MAT, 3
TYPE, 1
MSHKEY, 1
AMESH, 1, 2
AMESH, 4
AMESH, 7, 9
ALLSEL

MP, EX, 1, 65E9
MP, NUXY, 1, 0.25
MP, DENS, 1, 2400

MP, EX, 2, 20E9
MP, NUXY, 2, 0.3
MP, DENS, 2, 2400
```

Appendix 15

```
FPX=1.0D-10
ONE=1.0

TB, PM, 2, , , PERM              !MATERIAL MODEL FOR
                                  POROUS MEDIA
TBDATA, 1, FPX, FPX, FPX

MP, EX, 3, 1E9
MP, NUXY, 3, 0.25
MP, DENS, 3, 2400

NSEL, S, LOC, X, 400
NSEL, A, LOC, X, 0
D, ALL, UX, 0
ALLSEL

NSEL, S, LOC, Y, 0
D, ALL, UY, 0
ALLSEL

NSEL, S, LOC, Y, 125
SF, ALL, PRES, 100E6
ALLSEL

LSEL, S, LINE, , 17, 28
NSLL, S, 1
SF, ALL, PRES, 50E6
ALLSEL
FINISH
/SOLV
NLGEOM, ON
NSUBST, 100, 5000, 20
TIME, 1
KBC, 0
OUTRES, ALL, ALL
OUTRES, SVAR, ALL
ALLS
SOLVE
TIME, 1*3600*3
KBC, 1
NSUBST, 200, 1000, 50
LSEL, S, LINE, , 17, 28
NSLL, S, 1
D, ALL, PRES, 0
```

```
SF, ALL, PRES, 0
ALLSEL
SOLVE
FINI
```

Appendix 16: Input File of the Composite Damage Model in Section 19.3

```
/COM, download composite_damage.db from
/COM, www.feabea.net/models/composite_damage.db
/COM, UNITS (N, MM, MPA)
/PREP7
RESUME, 'composite_damage', 'db'
SECTYPE, 1, SHELL
SECDATA, 1.5, 1, 90
SECDATA, 1.5, 1, 0
SECDATA, 1.5, 1, 90
SECDATA, 1.5, 1, 0
SECDATA, 1.5, 1, 90
SECDATA, 1.5, 1, 0
SECDATA, 1.5, 1, 90
SECDATA, 1.5, 1, 0

SECTYPE, 2, SHELL
SECDATA, 1.5, 1, 0
SECDATA, 1.5, 1, 90
SECDATA, 1.5, 1, 0
SECDATA, 1.5, 1, 90
SECDATA, 1.5, 1, 0
SECDATA, 1.5, 1, 90
SECDATA, 1.5, 1, 0
SECDATA, 1.5, 1, 90

LOCAL, 11, 1, 0, 0, 0, 0, 0, 90
ESEL, S, TYPE, , 3, 5, 2
EMODIF, ALL, ESYS, 11
ALLSEL
CSYS, 0

!TYPICAL MATERIAL PROPERTIES FOR E-GLASS/EPOXY (MPA)
E_X=18.4*6894
E_Y=1.62*6894
E_Z=E_Y
```

```
PR_XY=0.1279
PR_YZ=0.1331
PR_XZ=PR_XY

G_XY=0.951*6894
G_YZ=0.528*6894
G_XZ=G_XY

MP, EX, 1, E_X
MP, EY, 1, E_Y
MP, EZ, 1, E_Z
MP, NUXY, 1, PR_XY
MP, NUYZ, 1, PR_YZ
MP, NUXZ, 1, PR_XZ
MP, GXY, 1, G_XY
MP, GYZ, 1, G_YZ
MP, GXZ, 1, G_XZ
MP, DENS, 1, 1
MP, DENS, 2, 1

TB, FCLI, 1, 1, , STRS        ! MATERIAL STRENGTHS
TBDATA, 1, 0.283*6894         !FAILURE STRESS,
                               X-DIRECTION TENSION
TBDATA, 2, -0.215*6894        !FAILURE STRESS, X-DIRECTION
                               COMPRESSION
TBDATA, 3, 6.96E-3*6894       !FAILURE STRESS, Y-DIRECTION
                               TENSION
TBDATA, 4, -2.90E-2*6894      !FAILURE STRESS, Y-DIRECTION
                               COMPRESSION
TBDATA, 5, 6.96E-3*6894       !FAILURE STRESS, Z-DIRECTION
                               TENSION !NOT USED BY HASHIN FC
TBDATA, 6, -2.90E-2*6894      !FAILURE STRESS, Z-DIRECTION
                    COMPRESSION !NOT USED BY HASHIN FC
TBDATA, 7, 1.15E-2*6894       !FAILURE STRESS,
                               XY-DIRECTION TENSION
TBDATA, 8, 7.25E-3*6894       !FAILURE STRESS,
                               YZ-DIRECTION TENSION
TBDATA, 9, 1.15E-2*6894       !FAILURE STRESS,
           XZ-DIRECTION TENSION !NOT USED BY HASHIN FC

TB, DMGI, 1, 1, , FCRT        ! FC FOR DAMAGE INITIATION
TBDATA, 1, 2, 2, 2, 2         ! MAX STRESS CRITERIA FOR
                               ALL FOUR FAILURE MODES
```

Appendix 16

```
TB, DMGE, 1, 1, , MPDG        ! DAMAGE EVOLUTION WITH MPDG
                                METHOD
TBTEMP, 0
TBDATA, 1, 0.6                ! 60% FIBER TENSION DAMAGE
                                (40% ULTIMATE STRENGTH)
TBDATA, 2, 0.6                ! 60% FIBER COMPRESSION
                              DAMAGE (40% ULTIMATE STRENGTH)
TBDATA, 3, 0.6                ! 60% MATRIX TENSION DAMAGE
                                (40% ULTIMATE STRENGTH)
TBDATA, 4, 0.6                ! 60% MATRIX COMPRESSION
                              DAMAGE (40% ULTIMATE STRENGTH)
MP, EX, 2, 18E4
MP, NUXY, 2, 0.3
MP, EX, 3, 180
MP, NUXY, 3, 0.286

ESEL, S, TYPE, , 1
EMODIF, ALL, MAT, 2
ALLSEL
ESEL, S, TYPE, , 4
EMODIF, ALL, MAT, 3
ALLSEL

NSEL, S, LOC, Z, 130
SF, ALL, PRES, 240
ALLSEL
NSEL, S, LOC, Z, 0
D, ALL, ALL
ALLSEL
FINISH

/SOLU
ANTYPE, 0
NLGEOM, ON
ERESX, NO
OUTRES, ALL, ALL
NSUB, 50, 10000, 20
TIME, 1
SOLV
FINISH


/POST26
NSOL, 3, 28284, U, Z,
PRVAR, 3, , , , , , , , ,
```

Appendix 17: Input File of the SLB Model in Section 19.4

```
/COM, UNITS (N, MM, MPA)
/COM, SLB PROBLEM TO VALIDATE CRACK GROWTH
/PREP7
DIS1=-5
N1=100
N2=100
N3=10
DL1=177.8
DL2=177.8-34.3
DH=2.032

ET, 1, 182              !* 2D 4-NODE STRUCTURAL SOLID
                              ELEMENT
KEYOPT, 1, 1, 2         !* ENHANCE STRAIN FORMULATION
KEYOPT, 1, 3, 2         !* PLANE STRAIN
ET, 2, 182
KEYOPT, 2, 1, 2
KEYOPT, 2, 3, 2
ET, 3, 202              ! 2D 4-NODE COHESIVE ZONE ELEMENT
KEYOPT, 3, 2, 2
KEYOPT, 3, 3, 2         ! PLANE STRAIN
MP, EX, 4, 135.3E3      ! ANISOTROPIC MATERIAL PROPERTIES
                              OF LAMINA
MP, EY, 4, 9.0E3
MP, EZ, 4, 9.0E3
MP, GXY, 4, 4.5E3
MP, GYZ, 4, 4.5E3
MP, GXZ, 4, 3.3E3
MP, PRXY, 4, 0.24
MP, PRXZ, 4, 0.24
MP, PRYZ, 4, 0.46

TB, CGCR, 1, , 3, LINEAR         ! LINEAR FRACTURE
CRITERION
TBDATA, 1, 0.05, 0.05, 0.05
```

```
TBLIST, ALL, ALL

RECTNG, 0, DL2, DH                  ! DEFINE AREAS
RECTNG, DL2, DL1, DH
RECTNG, 0, DL2, 0, -DH
LESIZE, ALL, DH/2
MAT, 4
MSHAPE, 0, 2D
AMESH, 1, 3, 2
AMESH, 2
CLOCAL, 12, , , , , 180
CSYS, 0
NSEL, S, LOC, X, 0, DL2-10
NUMMRG, NODES
ESLN
TYPE, 3
CZMESH, , , 1, Y, 0,                ! GENERATE INTERFACE
ELEMENTS
ALLSEL, ALL
ASEL, S, AREA, , 1, 2
NSLA, S, 1
NSEL, R, LOC, X, DL2
NUMMRG, NODES
ALLSEL

NSEL, S, LOC, X, 0                  ! APPLY CONSTRAINTS
NSEL, R, LOC, Y, 0
D, ALL, ALL
NSEL, ALL
NSEL, S, LOC, X, DL1
D, ALL, UY
ALLSEL
ESEL, S, ENAME, , 202
CM, CPATH1, ELEM

NSEL, S, LOC, X, DL2-10
NSEL, R, LOC, Y, 0
CM, CRACK1, NODE
ALLS
NSEL, S, LOC, X, DL1/2
NSEL, R, LOC, Y, DH                 ! APPLY DISPLACEMENT
LOADING ON TOP
D, ALL, UY, DIS1
```

Appendix 17

```
ALLSLE
FINISH

/SOLU
TIME, 1
CINT, NEW, 1
CINT, TYPE, VCCT
CINT, CTNC, CRACK1
CINT, NORM, 12, 2
CINT, NCON, 6            ! NUMBER OF COUNTOURS
CINT, SYMM, OFF

! CRACK GROWTH SIMULATION SET
CGROW, NEW, 1
CGROW, METHOD, VCCT
CGROW, CID, 1            ! CINT ID VCCT CALCULATION
CGROW, CPATH, CPATH1     ! CRACK PATH
CGROW, FCOP, MTAB, 1
CGROW, DTIME, 1E-3
CGROW, DTMIN, 1E-4
CGROW, DTMAX, 2E-2
NSUBST, N1, N2, N3
OUTRES, ALL, ALL
SOLV
FINISH

/POST1
SET, LAST
/GOPR
/OUT
PRCI, 1
PRCI, 1, , CEXT
/NOPR
/OUT, SCRATCH
/POST26
NTOP=NODE(DL1/2, DH, 0)
NSEL, ALL
NSOL, 2, NTOP, U, Y, UY
RFORCE, 3, NTOP, F, Y, FY
PROD, 4, 3, , , RF, , , 20
/TITLE, SLB: REACTION AT TOP NODE VERSUS PRESCRIBED
    DISPLACEMENT
/AXLAB, X, DISP U (MM)
```

```
/AXLAB, Y, REACTION FORCE R (N)
/YRANGE, 0, 60
XVAR, 2
PRVAR, UY, RF

/OUT
PRVAR, 2, 3, 4, 5
```

Appendix 18: Input File of the Spur Gear Model with FGM in Section 20.3

```
/COM, UNITS (N, MM, MPA)
/PREP7
K, 1,
K, 2, 14.1, -2.57
K, 3, 14.9, 1.57
K, 4, 27.4, 4.86
K, 5, 36.4, 9.85
K, 6, 36.4, 15.14
K, 7, 27.4, 20.43
K, 8, 14.75, 23.43
K, 9, 13.88, 27.71
K, 10, 0.125, 24.71
K, 11, 1.125, 12.14
K, 12, 16.5, 22.29
K, 13, 16.5, 2.71
BSPLIN, 6, 7, 12, 8
BSPLIN, 5, 4, 13, 3
BSPLIN, 10, 11, 1
LSTR, 8, 9
LSTR, 9, 10
LSTR, 3, 2
LSTR, 2, 1
LSTR, 6, 5
AL, 1, 2, 3, 4, 5, 6, 7, 8

ET, 1, 182
KEYOPT, 1, 3, 0

ESIZE, 1
AMESH, 1


TB, ELAS, 1                    ! DEFINITE THE
                                 MATERIAL PROPERTIES
*DO, I, 1, 11
 E%I%=2.01E5*EXP(0.0097*(I-1)*2)  ! ASSIGN YOUNG'S
                        MODULUS FOR THE SUPPORTING POINTS
```

```
  TBFIELD, XCOR,    (I-1)*2+14.9
  TBDATA,  1,      E%I% ,   0.3
*ENDDO
TBIN, ALGO

F, 1, FY, -1.1368/2
F, 3, FY, -1.1368/2
ALLSEL
LSEL, S, LINE, , 3, 7, 2
NSLL, S, 1
D, ALL, ALL
ALLSEL
FINISH

/SOLU
NSUBST, 1, 1, 1
SOLV
FINISH
```

Appendix 19: Input File of the Orthodontic Wire Model in Section 21.2

```
/COM, UNITS (N, MM, MPA)
/PREP7
ET, 1, 182
ET, 2, 185
K, 1, -0.318, 0.216
K, 2, 0.318, 0.216
K, 3, 0.318, -0.216
K, 4, -0.318, -0.216

K, 5, 0, 0
K, 6, 0, 0, 5.5
K, 7, 0, -2, 5.5
K, 67, 0, -2, 10
L, 5, 6
LARC, 6, 7, 67, 1
K, 8, 0, -2, 3
L, 7, 8
K, 9, 0, -3, 2
K, 89, 0, -3, 10
LARC, 8, 9, 89, 1
K, 10, 0, -5, 2
L, 9, 10
K, 11, 0, -6, 3
K, 101, 0, -6, 10
LARC, 10, 11, 101, 1
K, 12, 0, -6, 9.5
L, 11, 12

A, 1, 2, 3, 4
ESIZE, , 4
AMESH, 1
LSEL, S, LINE, , 1, 7
LESIZE, ALL, 0.2
VDRAG, 1, , , , , , 1, 2, 3, 4, 5, 6
VDRAG, 31, , , , , , 7
ACLEAR, 1
ALLSEL
```

```
MP, EX, 1, 5E4
MP, NUXY, 1, 0.3
!DEFINE SMA MATERIAL PROPERTIES
TB, SMA, 1, , , SUPE
TBDATA, 1, 500, 500, 300, 300, 0.07, 0.136

ALLSEL
NSEL, S, LOC, Z, 0
D, ALL, ALL
ALLSEL
FINISH

/SOLU
NLGEOM, ON
NSUBST, 50, 1000, 30
TIME, 1
D, NODE(0, -6, 9.5), UZ, 7
NALL
OUTRES, ALL, ALL
SOLVE
TIME, 2
D, NODE(0, -6, 9.5), UZ, 0
NALL
OUTRES, ALL, ALL
SOLVE
FINISH
```

Appendix 20: Input File of the Vacuum Tight Shape Memory Flange Model in Section 21.3

```
/COM, UNITS (N, MM, MPA)
/PREP7
! RADIUS OF THE ROD IS 85 MM
RR=84
K, 1, 0, 0
K, 2, 0, 900
K, 3, 85, 840
K, 4, 85, 0
K, 5, RR, 670
K, 6, RR+1, 370
K, 7, RR+25, 0
K, 8, RR+140, 0
K, 9, RR+140, 250
K, 10, RR+25, 250
K, 11, RR+15, 370
K, 12, RR+15, 670
A, 1, 2, 3, 4
A, 5, 6, 7, 8, 9, 10, 11, 12
K, 100, 0, 500
VROTAT, 1, 2, , , , , 1, 100, 90

ET, 1, 185

! DEFINE PROPERTIES FOR MATERIAL 1 (METAL)
MP, EX, 1, 53000
MP, NUXY, 1, 0.2

! DEFINE PROPERTIES FOR MATERIAL 2 (ROD)
MP, EX, 2, 74000
MP, NUXY, 2, 0.2

! DEFINE PROPERTIES FOR MATERIAL 5 (SMA)
MP, EX, 5, 60000
MP, NUXY, 5, 0.36
```

```
TB, SMA, 5, , , MEFF
TBDATA, 1, 1000, 223, 50, 2.1, 0.04, 45000, 0.0

ESIZE, 20
VSEL, S, VOLU, , 1
MAT, 2
VSWEEP, 1

VSEL, S, VOLU, , 2
MAT, 1
VSWEEP, 2
ALLSEL

ASEL, S, AREA, , 7
NSLA, S
ESLN, S
EMODIF, ALL, MAT, 5
ALLSEL
MAT, 3
R, 3, 3, 1
REAL, 3
ET, 2, 170
ET, 3, 174
KEYOPT, 3, 9, 1
KEYOPT, 3, 10, 2
R, 3,
! GENERATE THE TARGET SURFACE
ASEL, S, , , 4
CM, _TARGET, AREA
TYPE, 2
NSLA, S, 1
ESLN, S, 0
ESLL, U
ESEL, U, ENAME, , 188, 189
NSLE, A, CT2
ESURF
CMSEL, S, _ELEMCM
! GENERATE THE CONTACT SURFACE
ASEL, S, , , 7
CM, _CONTACT, AREA
TYPE, 3
NSLA, S, 1
ESLN, S, 0
NSLE, A, CT2
```

Appendix 20

```
ESURF
ALLSEL

NSEL, S, LOC, X, 0
D, ALL, UX,
NSEL, S, LOC, Z
D, ALL, UZ
ALLSEL

NSEL, S, LOC, Y, 0
D, ALL, UY
ALLSEL

ASEL, S, AREA, , 10
NSLA, S, 1
D, ALL, ALL
FINISH

/SOLVE
ANTYPE, STATIC
OUTRES, ALL, ALL
NROPT, UNSYM
KBC, 0
DELTIM, 0.001
SOLCONTROL, ON
AUTOTS, ON
TIME, 1
ESEL, S, MAT, , 3
EKILL, ALL              ! EKILL COMMAND
ALLSEL
SFA, 7, , PRES, 22
ALLSEL
BFUNIF, TEMP, 223
SOLVE

TIME, 2
DELTIM, 0.001
ESEL, S, MAT, , 3
EALIVE, ALL             ! EALIVE COMMAND
ALLSEL
SFA, 7, , PRES, 0
ALLSEL
BFUNIF, TEMP, 223
```

```
SOLVE

TIME, 3
BFUNIF, TEMP, 300
SOLVE
```

Appendix 21: Input File of the Piezoelectric Microaccelerometer Model in Section 22.4

```
/COM, UNITS (UN, UM, MPA)
/PREP7
/TITLE, SENSITIVITY UNDER 1G ACCELERATION
ET, 1, 186            ! SELECT ELEMENT TYPE FOR SI
                        MATERIALS
ET, 2, 226, 1001      ! SELECT ELEMENT TYPE FOR PZT
                        PIEZOELECTRONIC MATERIALS

!MATERIAL PROPERTIES OF THE MASS AND BEAM
MP, DENS, 1, 2.33E-15
MP, EX, 1, 1.9E+5
MP, NUXY, 1, 0.18
STRENGTH=4.2E+2

! **** PIEZOELECTRIC MATERIALS ****
MP, DENS, 2, 7.55E-15
```

! DEFINE DIELECTRIC CONSTANT [&] (PF/μM) ==> F/M = 1E6 PF/μM)

```
MP, PERX, 2, 10.443E-3     ! X-DIRECTION PERMITTIVITY
MP, PERY, 2, 10.443E-3     ! Y-DIRECTION PERMITTIVITY
MP, PERZ, 2, 6.4605E-3     ! Z-DIRECTION PERMITTIVITY
```

!PIEZOELECTRIC STIFFNESS (ELASTIC) COEFFICIENT MATRIX
[C] (μN/μM^2) ==> N/M^2 = 1E-6 μN/μM^2)

```
C11=11.425E+4
C12=5.8294E+4
C13=5.8525E+4
C22=11.425E+4
C23=5.8525E+4
C33=9.8181E+4
C44=2.0747E+4
C55=2.0747E+4
C66=2.6042E+4
TB, ANEL, 2                ! INPUT [C] MATRIX FOR PZT
```

```
TBDATA, 1, C11, C12, C13
TBDATA, 7, C22, C23
TBDATA, 12, C33
TBDATA, 16, C44
TBDATA, 19, C55
TBDATA, 21, C66

D31=-93E-6
D33=223E-6
D15=494E-6
E13=2.3687
E23=7.6006
E33=16.422
E51=10.249
TB, PIEZ, 2                    ! DEFINE PIEZOELECTRIC TABLE FOR PZT
TBDATA, 3, E13                 ! DEFINE MATRIX CONSTANTS
TBDATA, 6, E23
TBDATA, 9, E33
TBDATA, 14, E51

!GIVEN MASS DIMENSIONS
LMT=800                        ! MASS TOP LENGTH (µM)
HM=300                         ! MASS HEIGHT (µM)
LMB=LMT-HM*SQRT(2)             ! MASS BOTTOM LENGTH (µM)
H=5

LE=25                          ! LENGTH OF THE ELEMENT
BLE=20

! CREATE MASS VOLUME
K, 1, -LMT/2, -LMT/2, 0
K, 2, LMT/2, -LMT/2, 0
K, 3, LMT/2, LMT/2, 0
K, 4, -LMT/2, LMT/2, 0
K, 5, -LMB/2, -LMB/2, -HM
K, 6, LMB/2, -LMB/2, -HM
K, 7, LMB/2, LMB/2, -HM
K, 8, -LMB/2, LMB/2, -HM
V, 1, 2, 3, 4, 5, 6, 7, 8

TYPE, 1
MAT, 1
LSEL, S, LINE, , 5, 11, 2
LESIZE, ALL, , , 5
```

Appendix 21 295

```
LSEL, S, LINE, , 1, 4, 3
LESIZE, ALL, LE, , , , , , 0
ALLSEL
MSHAPE, 0, 3D
MSHKEY, 1
VMESH, 1

!CREATE 15-µM-THICKNESS MASS TOP VOLUME
K, 9, -LMT/2, -LMT/2, H      ! KEYPOINTS
K, 10, LMT/2, -LMT/2, H
K, 11, LMT/2, LMT/2, H
K, 12, -LMT/2, LMT/2, H
V, 1, 2, 3, 4, 9, 10, 11, 12, 13

TYPE, 1
MAT, 1
ESIZE, LE
MSHAPE, 0, 3D
MSHKEY, 1
VMESH, 2
ALLS

!GIVEN BEAM DIMENSIONS
L=400              ! LENGTH OF BEAM SUSPENSION (µM)
B=200              ! WIDTH OF BEAM SUSPENSION (µM)

!GIVEN PZT THIN FILM DIMENSION
SLPZT=3*L/10
HPZT=0.5           ! THICKNESS OF PZT THIN FILM (µM)
<--- MODIFY
D1=L/10
D2=L-D1-LPZT

APZT=LPZT*B                    !CALCULATE XY AREA OF A PZT
CPZT=(6.4605E-3)*(LPZT*B)/HPZT ! CALCULATE CAPACITANCE
OF A PZT: C = &*A/H

!CREATE BEAM
K, 13, LMT/2, -B/2, 0
K, 14, (L+LMT/2), -B/2, 0
K, 15, (L+LMT/2), B/2, 0
K, 16, LMT/2, B/2, 0
K, 17, LMT/2, -B/2, H
K, 18, (L+LMT/2), -B/2, H
K, 19, (L+LMT/2), B/2, H
```

```
K, 20, LMT/2, B/2, H
V, 13, 14, 15, 16, 17, 18, 19, 20

K, 21, B/2, LMT/2, 0
K, 22, B/2, L+LMT/2, 0
K, 23, -B/2, L+LMT/2, 0
K, 24, -B/2, LMT/2, 0
K, 25, B/2, LMT/2, H
K, 26, B/2, L+LMT/2, H
K, 27, -B/2, L+LMT/2, H
K, 28, -B/2, LMT/2, H
V, 21, 22, 23, 24, 25, 26, 27, 28

K, 29, -LMT/2, -B/2, 0
K, 30, -LMT/2, B/2, 0
K, 31, -(L+LMT/2), B/2, 0
K, 32, -(L+LMT/2), -B/2, 0
K, 33, -LMT/2, -B/2, H
K, 34, -LMT/2, B/2, H
K, 35, -(L+LMT/2), B/2, H
K, 36, -(L+LMT/2), -B/2, H
V, 29, 30, 31, 32, 33, 34, 35, 36

K, 37, B/2, -LMT/2, 0
K, 38, -B/2, -LMT/2, 0
K, 39, -B/2, -(L+LMT/2), 0
K, 40, B/2, -(L+LMT/2), 0
K, 41, B/2, -LMT/2, H
K, 42, -B/2, -LMT/2, H
K, 43, -B/2, -(L+LMT/2), H
K, 44, B/2, -(L+LMT/2), H
V, 37, 38, 39, 40, 41, 42, 43, 44

TYPE, 1
MAT, 1
ESIZE, BLE
LSEL, S, LINE, , 22
LSEL, A, LINE, , 24
LSEL, A, LINE, , 28
LSEL, A, LINE, , 32
LSEL, A, LINE, , 34
LSEL, A, LINE, , 36
LSEL, A, LINE, , 40
LSEL, A, LINE, , 44
LSEL, A, LINE, , 45
```

Appendix 21

```
LSEL, A, LINE, , 47
LSEL, A, LINE, , 50
LSEL, A, LINE, , 54
LSEL, A, LINE, , 57
LSEL, A, LINE, , 59
LSEL, A, LINE, , 62
LSEL, A, LINE, , 66
LESIZE, ALL, LE
MSHAPE, 0, 3D
MSHKEY, 1
VMESH, 3, 6, 1

! CREATE PZT THIN FILM VOLUME
K, 50, LMT/2+D1, -B/2, H
K, 51, LMT/2+D1+LPZT, -B/2, H
K, 52, LMT/2+D1+LPZT, B/2, H
K, 53, LMT/2+D1, B/2, H
K, 54, LMT/2+D1, -B/2, H+HPZT
K, 55, LMT/2+D1+LPZT, -B/2, H+HPZT
K, 56, LMT/2+D1+LPZT, B/2, H+HPZT
K, 57, LMT/2+D1, B/2, H+HPZT
V, 50, 51, 52, 53, 54, 55, 56, 57

K, 58, LMT/2+D2, -B/2, H
K, 59, LMT/2+D2+LPZT, -B/2, H
K, 60, LMT/2+D2+LPZT, B/2, H
K, 61, LMT/2+D2, B/2, H
K, 62, LMT/2+D2, -B/2, H+HPZT
K, 63, LMT/2+D2+LPZT, -B/2, H+HPZT
K, 64, LMT/2+D2+LPZT, B/2, H+HPZT
K, 65, LMT/2+D2, B/2, H+HPZT
V, 58, 59, 60, 61, 62, 63, 64, 65

K, 66, B/2, LMT/2+D1, H
K, 67, B/2, LMT/2+D1+LPZT, H
K, 68, -B/2, LMT/2+D1+LPZT, H
K, 69, -B/2, LMT/2+D1, H
K, 70, B/2, LMT/2+D1, H+HPZT
K, 71, B/2, LMT/2+D1+LPZT, H+HPZT
K, 72, -B/2, LMT/2+D1+LPZT, H+HPZT
K, 73, -B/2, LMT/2+D1, H+HPZT
V, 66, 67, 68, 69, 70, 71, 72, 73

K, 74, B/2, LMT/2+D2, H
K, 75, B/2, LMT/2+D2+LPZT, H
```

```
K, 76, -B/2, LMT/2+D2+LPZT, H
K, 77, -B/2, LMT/2+D2, H
K, 78, B/2, LMT/2+D2, H+HPZT
K, 79, B/2, LMT/2+D2+LPZT, H+HPZT
K, 80, -B/2, LMT/2+D2+LPZT, H+HPZT
K, 81, -B/2, LMT/2+D2, H+HPZT
V, 74, 75, 76, 77, 78, 79, 80, 81

K, 82, -(LMT/2+D1), -B/2, H
K, 83, -(LMT/2+D1), B/2, H
K, 84, -(LMT/2+D1+LPZT), B/2, H
K, 85, -(LMT/2+D1+LPZT), -B/2, H
K, 86, -(LMT/2+D1), -B/2, H+HPZT
K, 87, -(LMT/2+D1), B/2, H+HPZT
K, 88, -(LMT/2+D1+LPZT), B/2, H+HPZT
K, 89, -(LMT/2+D1+LPZT), -B/2, H+HPZT
V, 82, 83, 84, 85, 86, 87, 88, 89

K, 90, -(LMT/2+D2), -B/2, H
K, 91, -(LMT/2+D2), B/2, H
K, 92, -(LMT/2+D2+LPZT), B/2, H
K, 93, -(LMT/2+D2+LPZT), -B/2, H
K, 94, -(LMT/2+D2), -B/2, H+HPZT
K, 95, -(LMT/2+D2), B/2, H+HPZT
K, 96, -(LMT/2+D2+LPZT), B/2, H+HPZT
K, 97, -(LMT/2+D2+LPZT), -B/2, H+HPZT
V, 90, 91, 92, 93, 94, 95, 96, 97

K, 98, B/2, -(LMT/2+D1), H
K, 99, -B/2, -(LMT/2+D1), H
K, 100, -B/2, -(LMT/2+D1+LPZT), H
K, 101, B/2, -(LMT/2+D1+LPZT), H
K, 102, B/2, -(LMT/2+D1), H+HPZT
K, 103, -B/2, -(LMT/2+D1), H+HPZT
K, 104, -B/2, -(LMT/2+D1+LPZT), H+HPZT
K, 105, B/2, -(LMT/2+D1+LPZT), H+HPZT
V, 98, 99, 100, 101, 102, 103, 104, 105

K, 106, B/2, -(LMT/2+D2), H
K, 107, -B/2, -(LMT/2+D2), H
K, 108, -B/2, -(LMT/2+D2+LPZT), H
K, 109, B/2, -(LMT/2+D2+LPZT), H
K, 110, B/2, -(LMT/2+D2), H+HPZT
K, 111, -B/2, -(LMT/2+D2), H+HPZT
K, 112, -B/2, -(LMT/2+D2+LPZT), H+HPZT
```

Appendix 21

```
K, 113, B/2, -(LMT/2+D2+LPZT), H+HPZT
V, 106, 107, 108, 109, 110, 111, 112, 113
!
TYPE, 2
MAT, 2
ESIZE, BLE
VSEL, S, VOLU, , 7, 14
ASLV, S
LSLA, S
LSEL, R, LENGTH, , B
LESIZE, ALL, LE
LSEL, S, LENGTH, , HPZT
LESIZE, ALL, HPZT/2
MSHAPE, 0, 3D
MSHKEY, 1
VMESH, 7, 14, 1

NUMMRG, NODE, 1.0E-4    !MERGES COINCIDENT OR
                         EQUIVALENTLY DEFINED ITEMS
NUMCMP, NODE            !COMPRESSES THE NUMBERING OF
                         DEFINED ITEMS
NUMCMP, ELEM            !COMPRESSES THE NUMBERING OF
                         DEFINED ITEMS
NSEL, S, LOC, X, LMT/2+L
CM, D_NODE01, NODE
D, ALL, UX, 0
D, ALL, UY, 0
D, ALL, UZ, 0

NSEL, S, LOC, Y, LMT/2+L
CM, D_NODE02, NODE
D, ALL, UX, 0
D, ALL, UY, 0
D, ALL, UZ, 0

NSEL, S, LOC, X, -(LMT/2+L)
CM, D_NODE03, NODE
D, ALL, UX, 0
D, ALL, UY, 0
D, ALL, UZ, 0

NSEL, S, LOC, Y, -(LMT/2+L)
CM, D_NODE04, NODE
D, ALL, UX, 0
D, ALL, UY, 0
```

```
D, ALL, UZ, 0
ALLSEL

CLOCAL, 12, 0, , , , 90
ESYS, 12
VPLOT, 9, 10
VPLOT
VPLOT, 13, 14
VSEL, S, VOLU, , 9, 10
VSEL, A, VOLU, , 13, 14
VPLOT
ESLV, S
EMODIF, ALL, ESYS, 12
ALLSEL
FINISH

/SOLU
ANTYPE, 0
ALOAD=-9.81E+6
ESEL, S, TYPE, , 2
NSLE, S
NSEL, R, LOC, Z, 5
D, ALL, VOLT, 0
ALLSEL
OUTRES, ALL, ALL,
TIME, 1
NSUBST, 2, 100, 2
KBC, 0

ALLSEL
ACEL, , , ALOAD
SOLVE
FINISH

/POST1
SET, LAST
V11=VOLT(NODE(-760, -100, H+HPZT))+VOLT(NODE(-760, -75,
H+HPZT))+VOLT(NODE(-760, -50, H+HPZT))+VOLT(NODE(-760,
-25, H+HPZT))+VOLT(NODE(-760, 0, H+HPZT))
V12=VOLT(NODE(-760, 100, H+HPZT))+VOLT(NODE(-760, 75,
H+HPZT))+VOLT(NODE(-760, 50, H+HPZT))+VOLT(NODE(-760,
25, H+HPZT))
V21=VOLT(NODE(-640, -100, H+HPZT))+VOLT(NODE(-640, -75,
H+HPZT))+VOLT(NODE(-640, -50, H+HPZT))+VOLT(NODE(-640,
-25, H+HPZT))+VOLT(NODE(-640, 0, H+HPZT))
```

Appendix 21

```
V22=VOLT(NODE(-640, 100, H+HPZT))+VOLT(NODE(-640, 75,
H+HPZT))+VOLT(NODE(-640, 50, H+HPZT))+VOLT(NODE(-640,
25, H+HPZT))
DV1=(V11+V12-V21-V22)/5
V31=VOLT(NODE(-560, -100, H+HPZT))+VOLT(NODE(-560, -75,
H+HPZT))+VOLT(NODE(-560, -50, H+HPZT))+VOLT(NODE(-560,
-25, H+HPZT))+VOLT(NODE(-560, 0, H+HPZT))
V32=VOLT(NODE(-560, 100, H+HPZT))+VOLT(NODE(-560, 75,
H+HPZT))+VOLT(NODE(-560, 50, H+HPZT))+VOLT(NODE(-560,
25, H+HPZT))
V41=VOLT(NODE(-440, -100, H+HPZT))+VOLT(NODE(-440, -75,
H+HPZT))+VOLT(NODE(-440, -50, H+HPZT))+VOLT(NODE(-440,
-25, H+HPZT))+VOLT(NODE(-440, 0, H+HPZT))
V42=VOLT(NODE(-440, 100, H+HPZT))+VOLT(NODE(-440, 75,
H+HPZT))+VOLT(NODE(-440, 50, H+HPZT))+VOLT(NODE(-440,
25, H+HPZT))
DV2=(V31+V32-V41-V42)/5
DV1_2=DV1+DV2

V_OUTPUT=ABS(DV1_2)*4
/OUT,
*STATUS, V_OUTPUT
```

Appendix 22: Input File of the Contact Model in Section 23.2

```
/COM, download nano2.cdb from
/COM, www.feabea.net/models/nano2.cdb
/COM, UNITS (N, MM, MPA)
*DO, I, 1, 20

*IF, I, EQ, 1, THEN
CDREAD, DB, 'nano2', 'cdb', , '', ''
YM0=20E3
*ELSE
/CLEAR,
PARRES,
/PREP7
CDREAD, DB, 'NANO2', 'CDB', , '', ''
R5=3.9427*18.75-21.48
R6=3.9427*22.5-21.48
R7=3.9427*26.25-21.48
R8=3.9427*30-21.48
R5678=(R5+R6+R7+R8)/4*1E-6
YM=R5678/RFTOL*YM0
YM0=YM
MP, EX, I+10, YM
MP, NUXY, I+10, 0.3
ESEL, S, TYPE, , 7
EMODIF, ALL, MAT, I+10
ALLSEL
*ENDIF
FINISH

/SOLU
TIME, 1
NLGEOM, ON
NSUBST, 8, 8, 8
OUTRES, ALL, ALL
SOLV
FINISH
```

```
/POST1
RFTOL=0.0
*DO, K, 5, 8
SET, 1, K
NSEL, S, LOC, Y, 125E-6
*GET, NUM_N, NODE, 0, COUNT        ! GET NUMBER OF
                                     NODES
*GET, N_MIN, NODE, 0, NUM, MIN     ! GET MIN NODE
                                     NUMBER
RFTOL%K%=0.0
   *DO, J, 1, NUM_N, 1             ! OUTPUT TO ASCII
                                   BY LOOPING OVER NODES
      CURR_N=N_MIN
     *GET, RFN, NODE, CURR_N, RF, FY
      RFTOL%K%=RFTOL%K%+RFN
     *GET, N_MIN, NODE, CURR_N, NXTH
   *ENDDO
RFTOL=RFTOL+ABS(RFTOL%K%)
*ENDDO
*STATUS, RFTOL
*STATUS, YM
*STATUS, I
R5=3.9427*18.75-21.48
R6=3.9427*22.5-21.48
R7=3.9427*26.25-21.48
R8=3.9427*30-21.48
R5678=(R5+R6+R7+R8)/4*1E-6
*IF, RFTOL, LT, R5678*1.05, AND, RFTOL, GT, R5678*0.95, THEN
*STATUS, YM0
*EXIT
*ENDIF
PARSAV, SCALAR
*ENDDO
```

Index

Numbers & Symbols
π Plane, 119, 120, 129

A

ACEL, 75, 114, 133
Aggregate materials, 131
Anharmonicity, 7, 9
Anisotropic, 3, 139, 162, 171, 209, 212, 213, 216, 227, 228
ANSYS, 1–3, 5, 12, 13, 15–18, 20, 26, 27, 31, 37, 38, 47, 49, 59, 72–75, 77, 83, 89, 90, 101, 104, 109, 111–113, 119, 120, 127, 130, 131, 135, 136, 139–141, 146, 149, 150, 155, 157, 159, 161–165, 169, 171, 173, 175, 179, 186, 188, 189, 191, 207, 210, 212, 213, 219, 221, 223, 224, 229
ANSYS190, 23, 31, 36, 38, 53, 57, 67, 70, 84, 87, 98, 101, 113, 118, 120, 124, 131, 150, 163, 169, 173, 174, 191, 196, 207, 211
ANSYS elements
  CPT212, 150
  Plane182, 24, 31, 38, 69, 84, 113, 121
  SHELL181, 73, 163
  SOLID185, 91, 155, 163, 191, 196, 221
  SOLID187, 53, 73, 221
APDL, 54, 85, 91, 114
Atoms, 7, 8, 61, 219, 227
Austenite, 183, 184, 186, 188–190, 194, 196
Axial strain, 35, 36
Axial stress, 33, 35, 36
Axisymmetrical, 24, 28, 31–33, 113, 114, 118

B

Back stress, 37, 79
Bergstrom-Boyce model, 83, 84
Biaxial, 65
Bilinear isotropic hardening, 13–16
Bilinear kinematic hardening, 33
Bolt, 53–55, 57

Bond energy, 7, 8, 9
Breast surgery, 2, 59, 63, 228
Bulk forming, 23
Bulk modulus, 61, 63, 111

C

Cam Clay model, 2, 105, 109, 110–114, 118, 119, 228
  Extended Cam Clay model, 112
  modified Cam Clay model, 109, 110
Ceramics, 10
Chaboche model, 2, 19, 20, 21, 26, 27, 31, 33, 36–39, 45, 67, 229
Clay, 107, 111
Coefficient of friction, 121
Cohesion, 108, 119, 127, 131, 139
Combustion chamber, 2, 37–45
Composite materials, 159, 160–162, 175
Composites, 1, 2, 161, 162, 168, 170, 175, 227
Compression, 33, 63, 68–71, 84, 87, 91–93, 111, 127, 128, 130, 149, 187–191, 207, 221
Conical prism, 129
Consistency, 108
Consolidation, 2, 105, 149–151, 154, 155
Contact, 26, 85, 114, 115, 121, 145, 146, 196, 199, 200, 203, 221–223
Convergence pattern, 120, 124, 133
Covalent bonds, 61
Crack growth, 2, 157, 162, 170, 172–174
Creep, 2, 5, 47–49, 53, 55–57, 59, 79, 80, 229
Creep strain, 47, 49, 56
Curve-fitting, 16, 17, 19, 20, 21, 33, 63, 67–71, 229
Cutting-plane algorithm, 124, 136, 228
Cyclic loading, 2, 11, 31, 89

D

Damage, 10, 59, 89, 90–93, 95, 96, 98, 157, 161, 162, 165–170, 191
Damage accumulation, 2, 89

305

Damage evolution law, 162
Damage initiation criteria, 162
Darcy's Law, 149
Dashpots, 79, 80
Dies, 2, 23–26
Dilatancy angle, 131
Dipole moment, 208, 209
Drucker-Prager model, 2, 105, 119, 120, 124, 187, 228
Drucker's stability, 66
Ductility, 23, 61, 227

E

Eight-chain model, 65
EKILL, 142, 145
Elastic strain, 11, 20, 186, 210
Elastomers, 2, 59, 62–65, 89
Elastoplastic tangent, 15
Element birth and death, 145, 200
Element coordinate system, 164, 165, 169
Exponential Visco-Hardening (EVH), 26

F

Failure, 31, 47, 109, 127, 129, 142, 145, 161, 163, 169, 175
Fat, 73–75
Fatigue, 20, 159, 160
Ferroelectricity, 207
Finite element analysis, 1, 105, 157, 219
Finite element method, 73, 131, 141, 157, 211, 221, 223, 224
Flange, 2, 146, 157, 183, 196, 198, 199, 200, 204
Flow potential, 131, 139, 141
Forming process, 2, 23, 26, 27
Fracture, 172, 175
Frederick–Armstrong formulas, 19
Frictional angle, 119, 139
Functionally graded material (FGM), 157, 175–180

G

Glands, 73–75
Glass transition temperature, 62

H

Homogeneous, 74, 159
Hyperelasticity, 65, 101
    Arruda–Boyce model, 65–69, 71
    Gent model, 2, 64, 66, 101, 104
    Mooney–Rivlin model, 63–70, 73
    Neo-Hookean model, 63–66, 104
    Ogden model, 64, 66–70, 72, 85, 87
    Polynomial model, 64
    Yeoh model, 64, 65
Hysteresis loop, 31, 33

I

Inhomogeneous, 139
Initial stress, 112, 142, 146, 190
Initial yield surface, 109
Isotropic hardening, 1, 2, 11–18, 20, 21, 33

J

Joint, 124–127, 131
Jointed Rock material model, 136, 141, 145, 146, 228
Jointed Rock model, 2, 105, 135, 139, 140, 228

K

Kinematic hardening, 2, 11, 12, 19, 21, 31, 33, 36, 49

L

Large deformation, 2, 26, 59, 169, 184, 227
Least square method, 13
Local coordinate system, 164, 165, 169, 213

M

Macromolecule, 61, 62
Material parameters, 2, 5, 11–13, 19, 21, 26, 33, 39, 63, 64, 67, 68–71, 83–85, 101, 111, 114, 131, 132, 176, 187–189, 192, 193, 199, 223, 229
Matrix, 11, 111, 159, 209, 210, 212
Maximum virgin potential, 92

Index

Melting temperature, 62
Metal, 1–3, 5, 7–11, 19, 23, 31, 33, 37, 47, 61, 159, 161, 219
Metal forming, 5, 10, 23
Metallic bonds, 7, 8, 227
Microaccelerometer, 3, 157, 211, 217
Mode I, 172–174
Mode II, 172–174
Modern materials, 2, 3, 157, 225, 227
Mohr–Coulomb model, 2, 105, 120, 124, 127–133, 135, 136, 139, 140, 145, 146, 228
Mullins effect, 2, 59, 89, 90–93, 98
    Ogden–Roxburgh Mullins effect model, 89–91
    Qi–Boyce Mullins effect model, 89
Multilinear isotropic hardening model (MISO), 12

N

Nanomaterials, 219, 228
Nanometer, 219, 223
Nanoscale, 3, 219
Newton-Raphson iterations, 136
Non-associated plasticity, 130
Notched rod, 2, 31, 32

O

Orthodontic wire, 2, 157, 183, 191–196
Orthotropic, 162, 163, 210

P

Permeability, 107, 139, 149, 150
Perzyna option, 38
Piezoelectricity, 207, 209, 227
Piezoelectric materials, 3, 207–209
Plane strain, 38, 69, 84, 87, 121, 131, 135, 141, 145, 150, 154, 171
Plane stress, 176
Plastic deformation, 8, 31, 112, 131, 227
Plasticity, 1, 7, 10, 11, 23, 49, 109, 130
Plastic strain, 1, 11, 12, 15, 16, 20, 23, 27, 29, 33–35, 40–43, 111, 114, 117, 133
Plastic zone, 133
Poisson's ratio, 10, 61, 111, 112, 121, 150, 163, 212

Polymers, 2, 79
Pore pressure, 152–155, 227
Porosity, 107
Porous elasticity model, 111
Porous media, 149, 150
Prager-Lode type, 188
Pretension, 2, 53–55, 57
PZT, 211–217

Q

Quasi-static, 26

R

Ratcheting, 2, 5, 20, 31–33, 36
RATE, 26, 55, 56
Reaction force, 27, 30, 86, 87, 92, 93, 97, 98, 173, 221, 222
Rigid, 26, 69, 70, 85, 90–92, 98
Rubber tire, 2, 89–91
Rupture, 59, 108

S

Safety factor, 10, 23, 132–136
Sand, 107
Shape memory alloys (SMAs), 1, 2, 157, 183–186, 227
    austenite, 183
    detwinned martensite, 183, 184, 186
    martensite, 183–190, 194, 196
    phase transformation, 183, 184
    shape memory effect, 2, 185, 186, 188–190, 199
    superelasticity, 2, 183, 184, 186, 188, 189, 191–193, 195, 196
    transformation strains, 193
    twinned martensite, 183
Shearing stresses, 8
Shear modulus, 63, 111
Shear stress, 127, 129, 131, 139
Sheet metal forming, 23
Silt, 107
Simple shear, 66
Single-leg bending problem (SLB), 2
Skin, 73–76
Sleeve, 196, 199, 200
Slope, 2, 15, 112, 131–136, 190

Slope stability, 127, 131, 133, 136, 228
Softening, 2, 89, 109, 110, 131
Soft tissues, 2, 59, 79, 83, 84, 86
Soil, 1, 2, 3, 105, 107–109, 111, 112, 146, 225, 227, 228
Soil-arch interaction, 2, 119, 120, 121, 124
Solution setting, 1, 26, 75, 199, 200, 225, 228
Spur gear, 176–181
Standard contact, 26, 85, 114, 121
Standard Newton-Raphson method, 124
Stiffness, 11, 61, 84, 111, 118, 139, 149, 159, 209, 212, 227
Strain energy density function, 102
Stress concentration, 145, 176, 178, 179
Stress relaxation, 47, 48, 55, 80
Structural analysis, 1, 105, 113
Substeps, 26, 133, 136, 146, 169, 222, 228
Symmetrical condition, 31, 33, 40, 53, 54, 192, 215, 228

T

Tangent modulus, 33
TB, CREEP, 53–54, 55, 57
TBFIELD, 2
TBIN, ALGO, 175, 177, 179
TBTEMP, 37
Temperature-dependent, 2, 37
Tensile strength, 111, 131, 139, 140
Tension, 1, 8, 15, 18, 33, 66–68, 103, 130, 187–191, 207
Thermal strain, 7, 37, 40, 41, 44, 45, 200, 227
Thermoplasticity, 61
Time hardening, 47–49, 53
Total strain, 11, 15, 20, 186
Trial safety factors, 132–136
Tunnel excavation, 2, 105, 141, 142, 145, 146, 149, 228

U

Uniaxial, 8, 15, 18, 66–68, 70, 103, 119, 189
UserCreep, 49, 53–55, 57, 229
UserHyper, 59, 101–104, 229
User subroutine, 5, 47, 75, 225, 229

V

Van der Waals forces, 61
VCCT, 170, 172–174
Virgin material strain-energy potential, 92
Viscoelasticity, 2, 59, 79, 83–85, 87
  Burgers model, 80, 81
  Kelvin-Voigt model, 80, 81
  Maxwell model, 80–82
  Prony series, 82, 83, 85, 87
Voce hardening model, 16, 19
Voce law, 2, 16–18
Void ratio, 112
Voids, 107, 227
Von Mises stresses, 27, 29, 70, 86, 114, 122, 123, 142, 145, 152

W

Wear, 59

Y

Yield function, 11, 109, 110, 112, 113, 119, 187–189
Yield stress, 8, 15, 17, 37, 42, 119, 141
Yield surface, 11, 42, 109, 110, 111, 112, 119, 120, 124, 129, 130, 136, 139, 140, 228
Young's modulus, 3, 13, 17, 20, 37, 43, 61, 62, 113, 121, 150, 151, 157, 163, 177–180, 199, 212, 219, 221–224, 227